ALGEBRAIC NUMBERS AND FOURIER ANALYSIS

SELECTED PROBLEMS ON EXCEPTIONAL SETS

THE WADSWORTH MATHEMATICS SERIES

Series Editors
Raoul H. Bott, Harvard University
David Eisenbud, Brandeis University
Hugh L. Montgomery, University of Michigan
Paul J. Sally, Jr., University of Chicago
Barry Simon, California Institute of Technology
Richard P. Stanley, Massachusetts Institute of Technology

W. Beckner, A. Calderón, R. Fefferman, P. Jones, *Conference on Harmonic Analysis in Honor of Antoni Zygmund*
M. Behzad, G. Chartrand, L. Lesniak-Foster, *Graphs and Digraphs*
J. Cochran, *Applied Mathematics: Principles, Techniques, and Applications*
A. Garsia, *Topics in Almost Everywhere Convergence*
K. Stromberg, *An Introduction to Classical Real Analysis*
R. Salem, *Algebraic Numbers and Fourier Analysis,* and L. Carleson, *Selected Problems on Exceptional Sets*

ALGEBRAIC NUMBERS AND FOURIER ANALYSIS

RAPHAEL SALEM

SELECTED PROBLEMS ON EXCEPTIONAL SETS

LENNART CARLESON
MITTAG-LEFFLER INSTITUT

WADSWORTH INTERNATIONAL GROUP
Belmont, California
A Division of Wadsworth, Inc.

Mathematics Editor: John Kimmel
Production Editor: Diane Sipes

The text of *Algebraic Numbers and Fourier Analysis* has been reproduced from the original with no changes. Minor revisions have been made by the author to the text of *Selected Problems on Exceptional Sets*.

Printed in the United States of America

1 2 3 4 5 6 7 8 9 10—87 86 85 84 83

Library of Congress Cataloging in Publication Data

Salem, Raphaël.
Algebraic numbers and Fourier analysis.
(Wadsworth mathematics series)
Reprint. Originally published: Boston: Heath, 1963.
Reprint. Originally published: Princeton, N.J.: Van Nostrand, c1967.
Includes bibliographies and index.
1. Algebraic number theory. 2. Fourier analysis. 3. Harmonic analysis. 4. Potential, Theory of. I. Carleson, Lennart. Selected problems on exceptional sets. II. Title. III. Series.
QA247.S23 1983 512'.74 82-20053
ISBN 0-534-98049-X

ALGEBRAIC NUMBERS AND FOURIER ANALYSIS

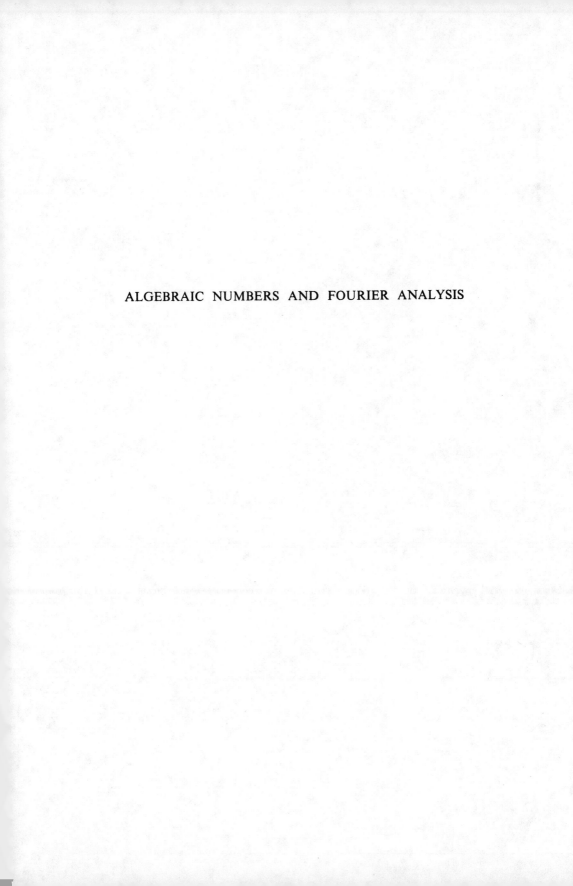

Algebraic Numbers and Fourier Analysis

RAPHAEL SALEM

To the memory of my father —

to the memory of my nephew, Emmanuel Amar,

who died in 1944 in a concentration camp —

to my wife and my children, to whom

I owe so much —

this book is dedicated

PREFACE

THIS SMALL BOOK contains, with but a few developments, the substance of the lectures I gave in the fall of 1960 at Brandeis University at the invitation of its Department of Mathematics.

Although some of the material contained in this book appears in the latest edition of Zygmund's treatise, the subject matter covered here has never until now been presented as a whole, and part of it has, in fact, appeared only in original memoirs. This, together with the presentation of a number of problems which remain unsolved, seems to justify a publication which, I hope, may be of some value to research students. In order to facilitate the reading of the book, I have included in an Appendix the definitions and the results (though elementary) borrowed from algebra and from number theory.

I wish to express my thanks to Dr. Abram L. Sachar, President of Brandeis University, and to the Department of Mathematics of the University for the invitation which allowed me to present this subject before a learned audience, as well as to Professor D. V. Widder, who has kindly suggested that I release my manuscript for publication in the series of *Heath Mathematical Monographs*. I am very grateful to Professor A. Zygmund and Professor J.-P. Kahane for having read carefully the manuscript, and for having made very useful suggestions.

R. Salem
Paris, 1 November 1961

Professor Raphaël Salem died suddenly in Paris on the twentieth of June, 1963, a few days after seeing final proof of his work.

CONTENTS

Chapter I

A REMARKABLE SET OF ALGEBRAIC INTEGERS

1. Introduction

We shall first recall some notation. Given any real number a, we shall denote by $[a]$ its integral part, that is, the integer such that

$$[a] \leq a < [a] + 1.$$

By (a) we shall denote the fractional part of a; that is,

$$[a] + (a) = a.$$

We shall denote by $\| a \|$ the absolute value of the difference between a and the nearest integer. Thus,

$$\| a \| = \min | a - n |, \quad n = 0, \pm 1, \pm 2, \ldots.$$

If m is the integer nearest to a, we shall also write

$$a = m + \{a\}$$

so that $\| a \|$ is the absolute value of $\{a\}$.

Next we consider a sequence of numbers † $u_1, u_2, \ldots, u_n, \ldots$ such that

$$0 \leq u_j < 1.$$

Let Δ be an interval contained in $(0, 1)$, and let $| \Delta |$ be its length. Suppose that among the first N members of the sequence there are $\nu(\Delta, N)$ numbers in the interval Δ. Then if for any fixed Δ we have

$$\lim_{N \to \infty} \frac{\nu(\Delta, N)}{N} = | \Delta |,$$

we say that the sequence $\{u_n\}$ is *uniformly distributed.* This means, roughly speaking, that each subinterval of $(0, 1)$ contains its proper quota of points.

We shall now extend this definition to the case where the numbers u_j do not fall between 0 and 1. For these we consider the fractional parts, (u_j), of u_j, and we say that the sequence $\{u_n\}$ is *uniformly distributed modulo 1* if the sequence of the fractional parts, $(u_1), (u_2), \ldots, (u_n), \ldots$, is uniformly distributed as defined above.

The notion of uniform distribution (which can be extended to several dimensions) is due to H. Weyl, who in a paper [16], ‡ by now classical, has also given a very useful criterion for determining whether a sequence is uniformly distributed modulo 1 (cf. Appendix, 7).

† By "number" we shall mean "real number" unless otherwise stated.

‡ See the Bibliography on page 67.

Without further investigation, we shall recall the following facts (see, for example, [2]).

1. If ξ is an irrational number, the sequence of the fractional parts $(n\xi)$, $n = 1, 2, \ldots$, is uniformly distributed. (This is obviously untrue for ξ rational.)
2. Let $P(x) = a_k x^k + \cdots + a_0$ be a polynomial where at least one coefficient a_j, with $j > 0$, is irrational. Then the sequence $P(n)$, $n = 1, 2, \ldots$, is uniformly distributed modulo 1.

The preceding results give us some information about the uniform distribution modulo 1 of numbers $f(n)$, $n = 1, 2, \ldots$, when $f(x)$ increases to ∞ with x not faster than a polynomial.

We also have some information on the behavior — from the viewpoint of uniform distribution — of functions $f(n)$ which increase to ∞ slower than n. We know, for instance, that the sequence an^α $(a > 0, 0 < \alpha < 1)$ is uniformly distributed modulo 1. The same is true for the sequence $a \log^\alpha n$ if $\alpha > 1$, but untrue if $\alpha < 1$.

However, almost nothing is known when the growth of $f(n)$ is exponential. Koksma [7] has proved that ω^n is uniformly distributed modulo 1 for *almost all* (in the Lebesgue sense) numbers $\omega > 1$, but nothing is known for particular values of ω. Thus, we do not know whether sequences as simple as e^n or $(\frac{3}{2})^n$ are or are not uniformly distributed modulo 1. We do not even know whether they are everywhere dense (modulo 1) on the interval (0, 1).

It is natural, then, to turn in the other direction and try to study the numbers $\omega > 1$ such that ω^n is "badly" distributed. Besides the case where ω is a rational integer (in which case for all n, ω^n is obviously congruent to 0 modulo 1), there are less trivial examples of distributions which are as far as possible from being uniform. Take, for example, the quadratic algebraic integer †

$$\omega = \tfrac{1}{2}(1 + \sqrt{5}) \text{ with conjugate } \tfrac{1}{2}(1 - \sqrt{5}) = \omega'.$$

Here $\omega^n + \omega'^n$ is a rational integer; that is,

$$\omega^n + \omega'^n \equiv 0 \pmod 1.$$

But $|\omega'| < 1$, and so $\omega'^n \to 0$ as $n \to \infty$, which means that $\omega^n \to 0$ (modulo 1). In other words, the sequence ω^n has (modulo 1) a single limit point, which is 0. This is a property shared by some other algebraic integers, as we shall see.

2. The algebraic integers of the class *S*

DEFINITION. *Let θ be an algebraic integer such that all its conjugates (not θ itself) have moduli strictly less than* 1. *Then we shall say that θ* belongs to the class *S*.‡

† For the convenience of the reader, some classical notions on algebraic integers are given in the Appendix.

‡ We shall always suppose (without loss of generality) that $\theta > 0$. θ is necessarily real. Although every natural integer belongs properly to *S*, it is convenient, to simplify many statements, to exclude the number 1 from *S*. Thus, in the definition we can always assume $\theta > 1$.

Then we have the following.

THEOREM 1. *If θ belongs to the class S, then θ^n tends to* 0 (*modulo* 1) *as $n \to \infty$.*

PROOF. Suppose that θ is of degree k and let $\alpha_1, \alpha_2, \ldots, \alpha_{k-1}$ be its conjugates. The number $\theta^n + \alpha_1^n + \cdots + \alpha_{k-1}^n$ is a rational integer. Since $|\alpha_j| < 1$ for all j, we have, denoting by ρ the greatest of the $|\alpha_j|$, $j = 1, 2, \ldots, k-1$,

$$|\alpha_1|^n + \cdots + |\alpha_{k-1}|^n < (k-1)\rho^n, \quad \rho < 1,$$

and thus, since $\theta^n + \alpha_1^n + \cdots + \alpha_{k-1}^n \equiv 0 \pmod 1$,

we see that (modulo 1) $\theta^n \to 0$, and even that it tends to zero in the same way as the general term of a convergent geometric progression.

With the notation of section 1, we write $\| \theta^n \| \to 0$.

Remark. The preceding result can be extended in the following way. Let λ be any algebraic integer of the field of θ, and let $\mu_1, \mu_2, \ldots, \mu_{k-1}$ be its conjugates. Then

$$\lambda\theta^n + \mu_1\alpha_1^n + \cdots + \mu_{k-1}\alpha_{k-1}^n$$

is again a rational integer, and thus $\| \lambda\theta^n \|$ also tends to zero as $n \to \infty$, as can be shown by an argument identical to the preceding one. Further generalizations are possible to other numbers λ.

Up to now, we have not constructed any number of the class S except the quadratic number $\frac{1}{2}(1 + \sqrt{5})$. (Of course, all rational integers belong trivially to S.) It will be of interest, therefore, to prove the following result [10].

THEOREM 2. *In every real algebraic field, there exist numbers of the class S.*†

PROOF. Denote by $\omega_1, \omega_2, \ldots, \omega_k$ a basis ‡ for the integers of the field, and let $\omega_1^{(i)}, \omega_2^{(i)}, \ldots, \omega_k^{(i)}$ for $i = 1, 2, \ldots, k-1$ be the numbers conjugate to $\omega_1, \omega_2, \ldots, \omega_k$. By Minkowski's theorem on linear forms [5] (cf. Appendix, 9), we can determine rational integers $x_1, x_2, \ldots, x_k$, not all zero, such that

$$|x_1\omega_1 + \cdots + x_k\omega_k| \leq A$$

$$|x_1\omega_1^{(i)} + \cdots + x_k\omega_k^{(i)}| \leq \rho < 1 \quad (i = 1, 2, \ldots, k-1)$$

provided

$$A\rho^{k-1} \geq \sqrt{|D|},$$

D being the discriminant of the field. For A large enough, this is always possible, and thus the integer of the field

$$\theta = x_1\omega_1 + \cdots + x_k\omega_k$$

belongs to the class S.

† We shall prove, more exactly, that there exist numbers of S having the degree of the field.

‡ The notion of "basis" of the integers of the field is not absolutely necessary for this proof, since we can take instead of $\omega_1, \ldots, \omega_k$ the numbers $1, \alpha, \ldots, \alpha^{h-1}$, where α is any integer of the field having the degree of the field.

3. Characterization of the numbers of the class S

The fundamental property of the numbers of the class S raises the following question.

Suppose that $\theta > 1$ is a number such that $\| \theta^n \| \to 0$ as $n \to \infty$ (or, more generally, that θ is such that there exists a real number λ such that $\| \lambda\theta^n \| \to 0$ as $n \to \infty$). Can we assert that θ is an algebraic integer belonging to the class S?

This important problem is still unsolved. But it can be answered positively if one of the two following conditions is satisfied in addition:

1. The sequence $\| \lambda\theta^n \|$ tends to zero rapidly enough to make the series $\sum \| \lambda\theta^n \|^2$ convergent.
2. We know beforehand that θ is algebraic.

In other words, we have the two following theorems.

THEOREM A. *If $\theta > 1$ is such that there exists a λ with*

$$\sum \| \lambda\theta^n \|^2 < \infty,$$

then θ is an algebraic integer of the class S, and λ is an algebraic number of the field of θ.

THEOREM B. *If $\theta > 1$ is an* algebraic number *such that there exists a real number λ with the property $\| \lambda\theta^n \| \to 0$ as $n \to \infty$, then θ is an algebraic integer of the class S, and λ is algebraic and belongs to the field of θ.*

The proof of Theorem A is based on several lemmas.

LEMMA I. *A necessary and sufficient condition for the power series*

$$f(z) = \sum_0^\infty c_n z^n \tag{1}$$

to represent a rational function,

$$\frac{P(z)}{Q(z)}$$

(P and Q polynomials), is that its coefficients satisfy a recurrence relation,

$$\alpha_0 c_m + \alpha_1 c_{m+1} + \cdots + \alpha_p c_{m+p} = 0,$$

valid for all $m \geq m_0$, the integer p and the coefficients $\alpha_0, \alpha_1, \ldots, \alpha_p$ being independent of m.

LEMMA II (Fatou's lemma). *If in the series (1) the coefficients c_n are rational integers and if the series represents a rational function, then*

$$f(z) = \frac{P(z)}{Q(z)},$$

where P/Q is irreducible, P and Q are polynomials with rational integral coefficients, and $Q(0) = 1$.

LEMMA III (Kronecker). *The series* (1) *represents a rational function if and only if the determinants*

$$\Delta_m = \begin{vmatrix} c_0 & c_1 & \dots & c_m \\ c_1 & c_2 & \dots & c_{m+1} \\ \dots & \dots & \dots & \dots \\ c_m & c_{m+1} & \dots & c_{2m} \end{vmatrix}$$

are all zero for $m \geq m_1$.

LEMMA IV (Hadamard). *Let the determinant*

$$D = \begin{vmatrix} a_1 & b_1 & \dots & l_1 \\ a_2 & b_2 & \dots & l_2 \\ \dots & \dots & \dots & \dots \\ a_n & b_n & \dots & l_n \end{vmatrix}$$

have real or complex elements. Then

$$| D^2 | \leq \left(\sum_1^n | a_j |^2\right)\left(\sum_1^n | b_j |^2\right)\cdots\left(\sum_1^n | l_j |^2\right).$$

We shall not prove here Lemma I, the proof of which is classical and almost immediate [3], nor Lemma IV, which can be found in all treatises on calculus [4]. We shall use Lemma IV only in the case where the elements of D are real; the proof in that case is much easier. For the convenience of the reader, we shall give the proofs of Lemma II and Lemma III.

PROOF of Lemma II. We start with a definition: A formal power series

$$\sum_0^\infty a_n z^n$$

with rational integral coefficients will be said to be *primitive* if no rational integer $d > 1$ exists which divides *all* coefficients.

Let us now show that if two series,

$$\sum_0^\infty a_n z^n \quad \text{and} \quad \sum_0^\infty b_n z^n,$$

are both primitive, their formal product,

$$\sum_0^\infty c_n z^n, \quad c_n = \sum_{\nu=0}^n a_\nu b_{n-\nu},$$

is also primitive. Suppose that the prime rational integer p divides *all* the c_n. Since p cannot divide all the a_n, suppose that

$$\left.\begin{array}{l} a_0 \equiv 0 \\ a_1 \equiv 0 \\ \dots\dots \\ a_{k-1} \equiv 0 \end{array}\right\} \pmod p, \quad a_k \not\equiv 0 \pmod p.$$

We should then have

$$\begin{aligned} c_k &\equiv a_k b_0 \pmod{p}, \text{ whence } b_0 \equiv 0 \pmod{p}, \\ c_{k+1} &\equiv a_k b_1 \pmod{p}, \text{ whence } b_1 \equiv 0 \pmod{p}, \\ c_{k+2} &\equiv a_k b_2 \pmod{p}, \text{ whence } b_2 \equiv 0 \pmod{p}, \end{aligned}$$

and so on, and thus

$$\sum_0^\infty b_n z^n$$

would not be primitive.

We now proceed to prove our lemma. Suppose that the coefficients c_n are rational integers, and that the series

$$\sum_0^\infty c_n z^n$$

represents a rational function

$$f(z) = \frac{P(z)}{Q(z)} = \frac{p_0 + p_1 z + \cdots + p_m z^m}{q_0 + q_1 z + \cdots + q_n z^n},$$

which we assume to be irreducible. As the polynomial $Q(z)$ is wholly determined (except for a constant factor), the equations

$$q_0 c_s + q_1 c_{s-1} + \cdots + q_n c_{s-n} = 0 \quad (s > m)$$

determine completely the coefficients q_j (except for a constant factor). Since the c_s are rational, there is a solution with all q_j rational integers, and it follows that the p_i are also rational integers.

We shall now prove that $q_0 = \pm 1$. One can assume that no integer $d > 1$ divides all p_i and all q_j. (Without loss of generality, we may suppose that there is no common divisor to all coefficients c_n; i.e., $\sum c_n z^n$ is primitive.) The polynomial Q is primitive, for otherwise if d divided q_j for all j, we should have

$$f\frac{Q}{d} = \frac{P}{d}$$

and d would divide all p_j, contrary to our hypothesis.

Now let U and V be polynomials with integral rational coefficients such that

$$PU + QV = m \neq 0,$$

m being an integer. Then

$$m = Q(Uf + V).$$

Since Q is primitive, $Uf + V$ cannot be primitive, for m is not primitive unless $|m| = 1$. Hence, the coefficients of $Uf + V$ are divisible by m. If γ_0 is the constant term of $Uf + V$, we have

$$m = q_0 \gamma_0,$$

and, thus, since m divides γ_0, one has $q_0 = \pm 1$, which proves Lemma II.

Proof of Lemma III. The recurrence relation of Lemma I,

$$\alpha_0 c_m + \alpha_1 c_{m+1} + \cdots + \alpha_p c_{m+p} = 0, \tag{2}$$

for all $m \geq m_0$, the integer p and the coefficients $\alpha_0, \ldots, \alpha_p$ being independent of m, shows that in the determinant

$$\Delta_m = \begin{vmatrix} c_0 & c_1 & \cdots & c_m \\ c_1 & c_2 & \cdots & c_{m+1} \\ \cdots & \cdots & \cdots & \cdots \\ c_m & c_{m+1} & \cdots & c_{2m} \end{vmatrix},$$

where $m \geq m_0 + p$, the columns of order $m_0, m_0 + 1, \ldots, m_0 + p$ are dependent; hence, $\Delta_m = 0$.

We must now show that if $\Delta_m = 0$ for $m \geq m_1$, then the c_n satisfy a recurrence relation of the type (2); if this is so, Lemma III follows from Lemma I. Let p be the first value of m for which $\Delta_m = 0$. Then the last column of Δ_p is a linear combination of the first p columns; that is:

$$L_{j+p} = \alpha_0 c_j + \alpha_1 c_{j+1} + \cdots + \alpha_{p-1} c_{j+p-1} + c_{j+p} = 0, \quad j = 0, 1, \ldots, p.$$

We shall now show that $L_{j+p} = 0$ for all values of j. Suppose that

$$L_{j+p} = 0, \quad j = 0, 1, 2, \ldots, m-1, \quad (m > p).$$

If we can prove that $L_{m+p} = 0$, we shall have proved our assertion by recurrence. Now let us write

$$\Delta_m = \left|\begin{array}{c|ccc} \Delta_{p-1} & c_p & \cdots & c_m \\ & \vdots & & \vdots \\ \hline c_p & & & c_{p+m} \\ \vdots & & \ddots & \vdots \\ c_m & c_{p+m} & \cdots & c_{2m} \end{array}\right|,$$

and let us add to every column of order $\geq p$ a linear combination with coefficients $\alpha_0, \alpha_1, \ldots, \alpha_{p-1}$ of the p preceding columns. Hence,

$$\Delta_m = \left|\begin{array}{c|ccc} \Delta_{p-1} & L_p & \cdots & L_m \\ & \vdots & & \vdots \\ \hline c_p & & & L_{p+m} \\ \vdots & & \ddots & \vdots \\ c_m & L_{p+m} & \cdots & L_{2m} \end{array}\right|,$$

and since the terms above the diagonal are all zero, we have

$$\Delta_m = (-1)^{p+m} \Delta_{p-1} (L_{p+m})^{m-p+1}.$$

Since $\Delta_m = 0$, we have $L_{m+p} = 0$, which we wanted to show, and Lemma III follows.

We can now prove Theorem A.

PROOF of Theorem A [10]. We write

$$\lambda\theta^n = a_n + \epsilon_n,$$

where a_n is a rational integer and $|\epsilon_n| \leq \frac{1}{2}$; thus $|\epsilon_n| = \|\lambda\theta^n\|$. Our hypothesis is, therefore, that the series $\sum \epsilon_n^2$ converges.

The first step will be to prove by application of Lemma III that the series

$$\sum_0^\infty a_n z^n$$

represents a rational function. Considering the determinant

$$\Delta_n = \begin{vmatrix} a_0 & a_1 & \cdots & a_n \\ a_1 & a_2 & \cdots & a_{n+1} \\ \cdots & \cdots & \cdots & \cdots \\ a_n & a_{n+1} & \cdots & a_{2n} \end{vmatrix},$$

we shall prove that $\Delta_n = 0$ for all n large enough. Writing

$$\eta_m = a_m - \theta a_{m-1} = \theta\epsilon_{m-1} - \epsilon_m,$$

we have

$$\eta_m^2 < (\theta^2 + 1)(\epsilon_{m-1}^2 + \epsilon_m^2).$$

Transforming the columns of Δ_n, beginning with the last one, we have

$$\Delta_n = \begin{vmatrix} a_0 & \eta_1 & \cdots & \eta_n \\ a_1 & \eta_2 & \cdots & \eta_{n+1} \\ \cdots & \cdots & \cdots & \cdots \\ a_n & \eta_{n+1} & \cdots & \eta_{2n} \end{vmatrix}$$

and, by Lemma IV,

$$\Delta_n^2 \leq \left(\sum_0^n a_m^2\right)\left(\sum_1^{n+1} \eta_m^2\right)\cdots\left(\sum_n^{2n} \eta_m^2\right)$$

$$\leq \left(\sum_0^n a_m^2\right) R_1 R_2 \cdots R_n$$

where R_h denotes the remainder of the convergent series

$$\sum_h^\infty \eta_m^2.$$

But, by the definition of a_m,

$$\sum_0^n a_m^2 < C\theta^{2n},$$

where $C = C(\lambda, \theta)$ depends on λ and θ only.

Hence,

$$\Delta_n^2 \leq C \prod_{h=1}^{n} (\theta^2 R_h),$$

and since $R_h \to 0$ for $h \to \infty$, $\Delta_n \to 0$ as $n \to \infty$, which proves, since Δ_n is a rational integer, that Δ_n is zero when n is larger than a certain integer.

Hence

$$\sum_0^\infty a_n z^n = \frac{P(z)}{Q(z)}, \quad \text{(irreducible)}$$

where, by Lemma III, P and Q are polynomials with rational integral coefficients and $Q(0) = 1$. Writing

$$Q(z) = 1 + q_1 z + \cdots + q_k z^k,$$

we have

$$\begin{aligned} f(z) &= \sum_0^\infty \epsilon_n z^n \\ &= \sum_0^\infty \lambda \theta^n z^n - \sum_0^\infty a_n z^n \\ &= \frac{\lambda}{1 - \theta z} - \frac{P(z)}{1 + q_1 z + \cdots + q_k z^k}. \end{aligned}$$

Since the radius of convergence of

$$\sum_0^\infty \epsilon_n z^n$$

is at least 1, we see that

$$Q(z) = 1 + q_1 z + \cdots + q_k z^k$$

has only one zero inside the unit circle, that is to say, $1/\theta$. Besides, since $\sum \epsilon_n^2 < \infty$, $f(z)$ has no pole of modulus 1;† hence, $Q(z)$ has one root, $1/\theta$, of modulus less than 1, all other roots being of modulus strictly larger than 1. The reciprocal polynomial,

$$z^k + q_1 z^{k-1} + \cdots + q_k,$$

has one root θ with modulus larger than 1, all other roots being strictly interior to the unit circle $|z| < 1$. Thus θ is, as stated, a number of the class S.

Since

$$-\frac{\lambda}{\theta} = \frac{P(1/\theta)}{Q'(1/\theta)},$$

λ is an algebraic number belonging to the field of θ.

† See footnote on page 10.

PROOF of Theorem B. In this theorem, we again write

$$\lambda\theta^n = a_n + \epsilon_n,$$

a_n being a rational integer and $|\epsilon_n| = \|\lambda\theta^n\| \leq \frac{1}{2}$. The assumption here is merely that $\epsilon_n \to 0$ as $n \to \infty$, without any hypothesis about the rapidity with which ϵ_n tends to zero. But here, we assume from the start that θ is algebraic, and we wish to prove that θ belongs to the class S.

Again, the first step will be to prove that the series

$$\sum_0^\infty a_n z^n$$

represents a rational function. But we shall not need here to make use of Lemma III. Let

$$A_0 + A_1\theta + \cdots + A_k\theta^k = 0$$

be the equation with rational integral coefficients which is satisfied by the algebraic number θ. We have, N being a positive integer,

$$\lambda\theta^N(A_0 + A_1\theta + \cdots + A_k\theta^k) = 0,$$

and, since

$$\lambda\theta^{N+p} = a_{N+p} + \epsilon_{N+p},$$

we have

$$A_0a_N + A_1a_{N+1} + \cdots + A_ka_{N+k} = -(A_0\epsilon_N + A_1\epsilon_{N+1} + \cdots + A_k\epsilon_{N+k}).$$

Since the A_j are fixed numbers, the second member tends to zero as $N \to \infty$, and since the first member is a rational integer, it follows that

$$A_0a_N + A_1a_{N+1} + \cdots + A_ka_{N+k} = 0$$

for all $N \geq N_0$. This is a recurrence relation satisfied by the coefficients a_n, and thus, by Lemma I, the series

$$\sum_0^\infty a_n z^n$$

represents a rational function.

From this point on, the proof follows identically the proof of Theorem A. (In order to show that $f(z)$ has no pole of modulus 1, the hypothesis $\epsilon_n \to 0$ is sufficient.†) Thus, the statement that θ belongs to the class S is proved.

† A power series $f(z) = \sum_0^\infty c_n z^n$ with $c_n = o(1)$ cannot have a pole *on* the unit circle. Suppose in fact, without loss of generality, that this pole is at the point $z = 1$. And let $z = r$ tend to $1 - 0$ along the real axis. Then $|f(z)| \leq \sum_0^\infty |c_n| r^n = o(1 - r)^{-1}$, which is impossible if $z = 1$ is a pole.

4. An unsolved problem

As we pointed out before stating Theorems A and B, if we know only that $\theta > 1$ is such that there exists a real λ with the condition $\| \lambda\theta^n \| \to 0$ as $n \to \infty$, we are unable to conclude that θ belongs to the class S. We are only able to draw this conclusion either if we know that $\sum \| \lambda\theta^n \|^2 < \infty$ or if we know that θ is algebraic. In other words, the problem that is open is the existence of *transcendental* numbers θ with the property $\| \lambda\theta^n \| \to 0$ as $n \to \infty$.

We shall prove here the only theorem known to us about the numbers θ such that there exists a λ with $\| \lambda\theta^n \| \to 0$ as $n \to \infty$ (without any further assumption).

THEOREM. *The set of all numbers θ having the preceding property is denumerable.*

PROOF. We again write

$$\lambda\theta^n = a_n + \epsilon_n$$

where a_n is an integer and $| \epsilon_n | = \| \lambda\theta^n \|$. We have

$$a_{n+2} - \frac{a_{n+1}^2}{a_n} = \frac{a_n a_{n+2} - a_{n+1}^2}{a_n}$$
$$= \frac{(\lambda\theta^n - \epsilon_n)(\lambda\theta^{n+2} - \epsilon_{n+2}) - (\lambda\theta^{n+1} - \epsilon_{n+1})^2}{\lambda\theta^n - \epsilon_n},$$

and an easy calculation shows that, since $\epsilon_n \to 0$, the last expression tends to zero as $n \to \infty$. Hence, for $n \geq n_0$, $n_0 = n_0(\lambda, \theta)$, we have

$$\left| a_{n+2} - \frac{a_{n+1}^2}{a_n} \right| < \frac{1}{2};$$

this shows that the integer a_{n+2} is uniquely determined by the two preceding integers, a_n, a_{n+1}. Hence, the infinite sequence of integers $\{a_n\}$ is determined uniquely by the first $n_0 + 1$ terms of the sequence.

This shows that the set of all possible sequences $\{a_n\}$ is denumerable, and, since

$$\theta = \lim \frac{a_{n+1}}{a_n},$$

that the set of all possible numbers θ is denumerable. The theorem is thus proved.

We can finally observe that since

$$\lambda = \lim \frac{a_n}{\theta^n},$$

the set of all values of λ is also denumerable.

EXERCISES

1. Let K be a real algebraic field of degree n. Let θ and θ' be two numbers of the class S, both of degree n and belonging to K. Then $\theta\theta'$ is a number of the class S. In particular, if q is any positive natural integer, θ^q belongs to S if θ does.

2. The result of Theorem A of this chapter can be improved in the sense that the hypothesis

$$\sum \| \lambda\theta^n \|^2 < \infty$$

can be replaced by the weaker one

$$\sum_{j=1}^{n} j \| \lambda\theta^j \|^2 = o(n).$$

It suffices, in the proof of Theorem A, and with the notations used in this proof, to remark that

$$(R_1 \cdots R_n)^{\frac{1}{n}} \leq \frac{R_1 + \cdots + R_n}{n}$$

and to show, by an easy calculation, that under the new hypothesis, the second member tends to zero for $n \to \infty$.

A PROPERTY OF THE SET OF NUMBERS OF THE CLASS S

1. The closure of the set of numbers belonging to S

THEOREM. *The set of numbers of the class S is a closed set.*

The proof of this theorem [12] is based on the following lemma.

LEMMA. *To every number θ of the class S there corresponds a real number λ such that $1 \leq \lambda < \theta$ and such that the series*

$$\sum_0^\infty \| \lambda\theta^n \|^2$$

converges with a sum less than an absolute constant (i.e., independent of θ and λ).

PROOF. Let $P(z)$ be the irreducible polynomial with rational integral coefficients having θ as one of its roots (all other roots being thus strictly interior to the unit circle $|z| < 1$), and write

$$P(z) = z^k + q_1 z^{k-1} + \cdots + q_k.$$

Let $Q(z)$ be the reciprocal polynomial

$$Q(z) = z^k P\left(\frac{1}{z}\right) = 1 + q_1 z + \cdots + q_k z^k.$$

We suppose first that P and Q are not identical, which amounts to supposing that θ is not a quadratic unit. (We shall revert later to this particular case.)

The power series

$$\frac{P(z)}{Q(z)} = c_0 + c_1 z + \cdots + c_n z^n + \cdots$$

has rational integral coefficients (since $Q(0) = 1$) and its radius of convergence is θ^{-1}. Let us determine μ such that

$$g(z) = \frac{\mu}{1 - \theta z} - \frac{P(z)}{Q(z)} \tag{1}$$

will be regular in the unit circle. If we set

$$P(z) = (z - \theta)P_1(z),$$
$$Q(z) = (1 - \theta z)Q_1(z),$$

then P_1 and Q_1 are reciprocal polynomials, and we have

$$\mu = \left(\frac{1}{\theta} - \theta\right) \frac{P_1(1/\theta)}{Q_1(1/\theta)}.$$

Since $\left|\dfrac{P_1(z)}{Q_1(z)}\right| = 1$ for $|z| = 1$, and since $\dfrac{P_1}{Q_1}$ is regular for $|z| \leq 1$, we have

$$\left|\frac{P_1(\theta^{-1})}{Q_1(\theta^{-1})}\right| < 1,$$

and, thus,

$$|\mu| < \theta - \frac{1}{\theta} < \theta. \tag{2}$$

Finally,

$$g(z) = \sum_0^\infty \mu\theta^n z^n - \sum_0^\infty c_n z^n$$
$$= \sum_0^\infty (\mu\theta^n - c_n) z^n$$

has a radius of convergence *larger* than 1, since the roots of $Q(z)$ different from θ^{-1} are all *exterior* to the unit circle. Hence,

$$\sum_0^\infty (\mu\theta^n - c_n)^2 = \frac{1}{2\pi}\int_0^{2\pi} |g(e^{i\varphi})|^2 \, d\varphi.$$

But, by (1) and (2), we have for $|z| = 1$

$$|g(z)| < \frac{|\mu|}{\theta - 1} + \left|\frac{P}{Q}\right| < \frac{\theta^2 - 1}{\theta(\theta - 1)} + 1 = 2 + \frac{1}{\theta} < 3.$$

Hence,

$$\sum_0^\infty (\mu\theta^n - c_n)^2 < 9$$

which, of course, gives

$$\sum_0^\infty \| \mu\theta^n \|^2 < 9. \tag{3}$$

Now, by (2) $|\mu| < \theta$ and one can assume, by changing, if necessary, the sign of $\dfrac{P}{Q}$, that $\mu > 0$. (The case $\mu = 0$, which would imply $P\left(\dfrac{1}{\theta}\right) = 0$, is excluded for the moment, since we have assumed that θ is not a quadratic unit.) We can, therefore, write $0 < \mu < \theta$.

To finish the proof of the lemma, we suppose $\mu < 1$. (Otherwise we can take $\lambda = \mu$ and there is nothing to prove.) There exists an integer s such that

$$\frac{1}{\theta^s} \leq \mu < \frac{1}{\theta^{s-1}}$$

or

$$1 \leq \theta^s \mu < \theta,$$

We take $\lambda = \theta^s \mu$ and have by (3)

$$\sum_0^\infty \| \lambda\theta^n \|^2 = \sum_0^\infty \| \mu\theta^{n+s} \|^2$$

$$= \sum_s^\infty \| \mu\theta^m \|^2$$

$$< \sum_0^\infty \| \mu\theta^m \|^2 < 9.$$

Since $1 \leq \lambda < \theta$, this last inequality proves the lemma when θ is not a quadratic unit.

It remains to consider the case when θ is a quadratic unit. (This particular case is not necessary for the proof of the theorem, but we give it for the sake of completeness.) In this case

$$\theta^n + \theta^{-n}$$

is a rational integer, and

$$\| \theta^n \| \leq \frac{1}{\theta^n}.$$

Thus,

$$\sum_0^\infty \| \theta^n \|^2 < \sum_0^\infty \frac{1}{\theta^{2n}} = \frac{\theta^2}{\theta^2 - 1}$$

and since $\theta + \dfrac{1}{\theta}$ is at least equal to 3, we have $\theta \geq 2$ and

$$\frac{\theta^2}{\theta^2 - 1} < \frac{4}{3}.$$

Thus, since $\sum \| \theta^n \|^2 < \frac{4}{3}$, the lemma remains true, with $\lambda = 1$.

Remark. Instead of considering in the lemma the convergence of

$$\sum_0^\infty \| \lambda\theta^n \|^2$$

we can consider the convergence (obviously equivalent) of

$$\sum_0^\infty \sin^2 \pi\lambda\theta^n.$$

In this case we have

$$\sum_0^\infty \sin^2 \pi\lambda\theta^n \leq 9\pi^2.$$

PROOF of the theorem. Consider a sequence of numbers of the class S, $\theta_1, \theta_2, \ldots, \theta_p, \ldots$ tending to a number ω. We have to prove that ω belongs to S also.

Let us associate to every θ_p the corresponding λ_p of the lemma such that

$$(4) \qquad 1 \leq \lambda_p < \theta_p, \quad \sum_0^\infty \sin^2 \pi\lambda_p\theta_p{}^n < 9\pi^2.$$

Considering, if necessary, a subsequence only of the θ_p, we can assume that the λ_p which are included, for p large enough, between 1 and, say, 2ω, tend to a limit μ. Then (4) gives immediately

$$\sum_0^\infty \sin^2 \pi\mu\omega^n \leq 9\pi^2$$

which, by Theorem A of Chapter I, proves that ω belongs to the class S. Hence, the set of all numbers of S is closed.

It follows that 1 is not a limit point of S. In fact it is immediate that $\theta \in S$ implies, for all integers $q > 0$, that $\theta^q \in S$. Hence, if $1 + \epsilon_m \in S$, with $\epsilon_m \to 0$, one would have

$$(1 + \epsilon_m)^{\left[\frac{\alpha}{\epsilon_m}\right]} \in S,$$

α being any real positive number and $\left[\frac{\alpha}{\epsilon_m}\right]$ denoting the integral part of $\frac{\alpha}{\epsilon_m}$. But, as $m \to \infty$, $\epsilon_m \to 0$ and

$$(1 + \epsilon_m)^{\left[\frac{\alpha}{\epsilon_m}\right]} \to e^\alpha.$$

It would follow that the numbers of S would be everywhere dense, which is contrary to our theorem.

2. Another proof of the closure of the set of numbers belonging to the class S

This proof, [13], [11], is interesting because it may be applicable to different problems.

Let us first recall a classical definition: If $f(z)$ is analytic and regular in the unit circle $|z| < 1$, we say that it *belongs to the class* H^p $(p > 0)$ if the integral

$$\int_0^{2\pi} |f(re^{i\varphi})|^p \, d\varphi \quad (r < 1)$$

is bounded for $r < 1$. (See, e.g., [17].)

This definition can be extended in the following way. Suppose that $f(z)$ is meromorphic for $|z| < 1$, and that it has only a finite number of poles there (nothing is assumed for $|z| = 1$). Let $z_1, \ldots, z_m$, be the poles and denote by $P_j(z)$ the principal part of $f(z)$ in the neighborhood of z_j. Then the function

$$g(z) = f(z) - \sum_{j=1}^m P_j(z)$$

is regular for $|z| < 1$, and if $g(z) \in H^p$ (in the classical sense), we shall say that $f(z) \in H^p$ (in the extended sense).

We can now state Theorem A of Chapter I in the following equivalent form.

THEOREM A′. *Let $f(z)$ be analytic, regular in the neighborhood of the origin, and such that its expansion there*

$$\sum_0^\infty a_n z^n$$

has rational integral coefficients. Suppose that $f(z)$ is regular for $|z| < 1$, except for a simple pole $1/\theta$ $(\theta > 1)$. Then, if $f(z) \in H^2$, it is a rational function and θ belongs to the class S.

The reader will see at once that the two forms of Theorem A are equivalent.

Now, before giving the new proof of the theorem of the closure of S, we shall prove a lemma.

LEMMA. *Let $P(z)$ be the irreducible polynomial having rational integral coefficients and having a number $\theta \in S$ for one of its roots. Let*

$$Q(z) = z^k P\left(\frac{1}{z}\right)$$

be the reciprocal polynomial (k being the degree of P). Let λ be such that

$$\frac{\lambda}{1-\theta z} - \frac{P(z)}{Q(z)}$$

is regular in the neighborhood of $1/\theta$ and, hence, for all $|z| < 1$. [We have already seen that $|\lambda| < \theta - \dfrac{1}{\theta}$ (and that thus, changing if necessary the sign of Q, we can take $0 < \lambda < \theta - \dfrac{1}{\theta}$).] *Then, in the opposite direction* [11],

$$\lambda > \frac{1}{2(\theta+1)},$$

provided θ is not quadratic, and thus $P \neq Q$.

PROOF. We have already seen that

$$\frac{P(z)}{Q(z)} = \sum_0^\infty c_n z^n,$$

the coefficients c_n being rational integers. We now write

$$g(z) = \frac{\lambda}{1-\theta z} - \frac{P(z)}{Q(z)} = \sum_0^\infty (\lambda\theta^n - c_n) z^n = \sum_0^\infty \epsilon_n z^n.$$

We have

$$I = \frac{1}{2\pi}\int_0^{2\pi} |g(e^{i\varphi})|^2\, d\varphi = \sum_0^\infty \epsilon_n^2, \tag{5}$$

as already stated.

On the other hand, the integral can be written

$$I = \frac{1}{2\pi i}\int_{C} \left\{\frac{\lambda}{1-\theta z} - \frac{P}{Q}(z)\right\}\left\{\frac{\lambda}{1-\dfrac{\theta}{z}} - \frac{P}{Q}\left(\frac{1}{z}\right)\right\}\frac{dz}{z},$$

where the integral is taken along the unit circle, or

$$I = \frac{1}{2\pi i}\int_{C} \left\{\frac{\lambda}{1-\theta z} - \frac{P}{Q}\right\}\left\{\frac{\lambda}{z-\theta} - \frac{Q}{zP}\right\} dz.$$

But changing z into $1/z$, we have

$$\int_{C} \frac{\lambda}{1-\theta z}\frac{Q}{zP}\,dz = -\int_{C} \frac{\lambda}{1-\dfrac{\theta}{z}}\frac{P}{Q}\frac{dz}{z} = \int_{C} \frac{\lambda}{z-\theta}\frac{P}{Q}\,dz$$

$$= \frac{\lambda}{\dfrac{1}{\theta}-\theta}\lim\left\{\left(z-\frac{1}{\theta}\right)\frac{P}{Q}\right\}_{z=\frac{1}{\theta}} = \frac{\lambda}{\dfrac{1}{\theta}-\theta}\left(-\frac{\lambda}{\theta}\right) = \frac{\lambda^2}{\theta^2-1}.$$

Therefore,

$$I = \frac{1}{2\pi i}\int_{C} \frac{dz}{z} - \frac{2\lambda^2}{\theta^2-1} + \frac{1}{2\pi i}\int_{C} \frac{\lambda^2}{(1-\theta z)(z-\theta)}\,dz$$

$$= 1 - \frac{2\lambda^2}{\theta^2-1} + \frac{\lambda^2}{\dfrac{1}{\theta}-\theta}\left[\frac{z-\dfrac{1}{\theta}}{1-\theta z}\right]_{z=\frac{1}{\theta}}$$

$$= 1 - \frac{2\lambda^2}{\theta^2-1} + \frac{\lambda^2}{\theta^2-1} = 1 - \frac{\lambda^2}{\theta^2-1}$$

and thus (5) gives

$$1 - \frac{\lambda^2}{\theta^2-1} = \sum_{0}^{\infty} \epsilon_n^2. \tag{6}$$

This leads to

$$|\lambda| < \sqrt{\theta^2-1}$$

or changing, if necessary, the sign of Q, to $\lambda < \sqrt{\theta^2-1}$ (an inequality weaker than $\lambda < \theta - \dfrac{1}{\theta}$ already obtained in (2)).

On the other hand, since $\lambda - c_0 = \epsilon_0$, we have

$$|\lambda - c_0| = |\epsilon_0| < 1.$$

But

$$c_0 = \frac{P(0)}{Q(0)} = \frac{q_k}{\pm 1} \geq 1.$$

Hence $\lambda > 0$ and $c_0 < \lambda + 1$.

We shall now prove that

$$\lambda > \frac{1}{2(\theta + 1)}.$$

In fact, suppose that

$$\lambda \le \frac{1}{2(\theta + 1)};$$

then $\lambda < \frac{1}{4}$ and necessarily $c_0 = 1$. But, since

$$\frac{P}{Q}(1 - \theta z) = c_0 + \sum_1^\infty (c_n - \theta c_{n-1}) z^n,$$

we have, if $z = e^{i\varphi}$,

$$\frac{1}{2\pi} \int_0^{2\pi} \left| \frac{P}{Q}(1 - \theta z) \right|^2 d\varphi = c_0^2 + \sum_1^\infty (c_n - \theta c_{n-1})^2,$$

and since $\left| \frac{P}{Q} \right| = 1$ for $|z| = 1$ and the integral is

$$1 + \theta^2,$$

the equality $c_0 = 1$ implies

$$|c_1 - \theta| < \theta.$$

Hence, since c_1 is an integer, $c_1 \ge 1$.

And thus, since by (6)

$$\frac{\lambda^2}{\theta^2 - 1} + \epsilon_0^2 + \epsilon_1^2 < 1,$$

we have, with $c_0 = 1$, $c_1 \ge 1$, $\lambda\theta \le \frac{1}{2}$,

$$\frac{\lambda^2}{\theta^2 - 1} + (\lambda - 1)^2 + (\lambda\theta - 1)^2 < \frac{\lambda^2}{\theta^2 - 1} + (\lambda - c_0)^2 + (\lambda\theta - c_1)^2 < 1,$$

$$\frac{\lambda^2}{\theta^2 - 1} + \lambda^2(1 + \theta^2) - 2\lambda(1 + \theta) + 1 < 0,$$

$$\frac{\lambda^2\theta^4}{\theta^2 - 1} - 2\lambda(1 + \theta) + 1 < 0.$$

This contradicts

$$\lambda \le \frac{1}{2(\theta + 1)}.$$

Thus, as stated,

$$\lambda > \frac{1}{2(1 + \theta)}.$$

We can now give the new proof of the theorem stating the closure of S.

PROOF. Let ω be a limit point of the set S, and suppose first $\omega > 1$. Let $\{\theta_s\}$ be an infinite sequence of numbers of S, tending to ω as $s \to \infty$. Denote by $P_s(z)$ the irreducible polynomial with rational integral coefficients and having the root θ_s and let K_s be its degree (the coefficient of z^{K_s} being 1). Let

$$Q_s(z) = z^{K_s} P_s\left(\frac{1}{z}\right)$$

be the reciprocal polynomial. The rational function P_s/Q_s is regular for $|z| \leq 1$ except for a single pole at $z = \theta_s^{-1}$, and its expansion around the origin

$$\frac{P_s}{Q_s} = \sum_{n=0}^{\infty} a_n^{(s)} z^n$$

has rational integral coefficients.

Determine now λ_s such that

$$g_s(z) = \frac{\lambda_s}{1 - \theta_s z} - \frac{P_s(z)}{Q_s(z)} \tag{7}$$

will be regular for $|z| \leq 1$. (We can discard in the sequence $\{\theta_s\}$ the quadratic units, for since $\theta_s \to \omega$, K_s is necessarily unbounded.)† By the lemma, and changing, if necessary, the sign of Q_s, we have

$$\frac{1}{2(\theta_s + 1)} < \lambda_s < \theta_s.$$

Therefore, we can extract from the sequence $\{\lambda_s\}$ a subsequence tending to a limit different from 0. (We avoid complicating the notations by assuming that this subsequence is the original sequence itself.)

On the other hand, if $|z| = 1$,

$$|g_s(z)| < \frac{|\lambda_s|}{\theta_s - 1} + 1 < A,$$

A being a constant independent of s. Since $g_s(z)$ is regular, this inequality holds for $|z| \leq 1$.

We can then extract from the sequence $\{g_s(z)\}$, which forms a normal family, a subsequence tending to a limit $g^*(z)$. (And again we suppose, as we may, that this subsequence is the original sequence itself.) Then (7) gives

$$g^*(z) = \frac{\mu}{1 - \omega z} - \lim \frac{P_s}{Q_s}.$$

Since the coefficients $a_n^{(s)}$ of the expansion of P_s/Q_s are rational integers, their limits can only be rational integers. Thus the limit of P_s/Q_s satisfies all requirements of Theorem A'. (The fact that $g^*(z) \in H^2$ is a trivial consequence of its

† See Appendix, 5.

boundedness, since $| g^*(z) | \leq A$.) Therefore ω is a number of the class S, since $1/\omega$ is actually a pole for

$$\lim \frac{P_s}{Q_s},$$

because $\mu \neq 0$. (This is essential, and is the reason for proving a lemma to the effect that the λ_s are bounded *below*.)

EXERCISE

Let a be a natural positive integer ≥ 2. Then a is a limit point for the numbers of the class S. (Considering the equation

$$z^n(z - a) - 1 = 0,$$

the result for $a > 2$ is a straightforward application of Rouché's theorem. With a little care, the argument can be extended to $a = 2$.)

APPLICATIONS TO THE THEORY OF POWER SERIES; ANOTHER CLASS OF ALGEBRAIC INTEGERS

1. A generalization of the preceding results

Theorem A′ of Chapter II can be extended, and thus restated in the following way.

THEOREM A″. *Let $f(z)$ be analytic, regular in the neighborhood of the origin, and such that the coefficients of its expansion in this neighborhood,*

$$\sum_0^\infty a_n z^n$$

are either rational integers or integers of an imaginary quadratic field. Suppose that $f(z)$ is regular for $|z| < 1$ except for a finite number of poles $1/\theta_i$ ($|\theta_i| > 1$, $i = 1, 2, \ldots, k$). Then if $f(z)$ belongs to the class H^2 (in the extended sense), $f(z)$ is a rational function, and the θ_i are algebraic integers.

The new features of this theorem, when compared with Theorem A′, are:

1. We can have several (although a *finite* number of) poles.
2. The coefficients a_n need not be rational integers, but can be integers of an imaginary quadratic field.

Nevertheless, the proof, like that for Theorem A′, follows exactly the pattern of the proof of Theorem A (see [10]). Everything depends on showing that a certain Kronecker determinant is zero when its order is large enough. The transformation of the determinant is based on the same idea, and the fact that it is zero is proved by showing that it tends to zero. For this purpose, one uses the well-known fact [9] that the integers of imaginary quadratic fields share with the rational integers the property of not having zero as a limit point.

Theorem A″ shows, in particular, that if

$$f(z) = \sum_0^\infty a_n z^n,$$

where the a_n are rational integers, is regular in the neighborhood of $z = 0$, has only a finite number of poles in $|z| < 1$, and is uniformly bounded in the neighborhood of the circumference $|z| = 1$, then $f(z)$ is a rational function.

This result suggests the following extension.

THEOREM I. *Let*

$$f(z) = \sum_0^\infty a_n z^n,$$

where the a_n are rational integers, be regular in the neighborhood of $z = 0$, and *suppose that $f(z)$ is regular for $|z| < 1$ except for a finite number of poles. Let α be any imaginary or real number. If there exist two positive numbers, δ, η ($\eta < 1$) such that $|f(z) - \alpha| > \delta$ for $1 - \eta \leq |z| < 1$, then $f(z)$ is a rational function.*

PROOF. For the sake of simplicity, we shall assume that there is only *one* pole, the proof in this case being typical. We shall also suppose, to begin with, that $\alpha = 0$, and we shall revert later to the general case.

Let ϵ be any positive number such that $\epsilon < \eta$. If ϵ is small enough, there is *one* pole of $f(z)$ for $|z| < 1 - \epsilon$, and, say N zeros, N being independent of ϵ. Consider

$$g(z) = \frac{1}{1 + mz\, f(z)},$$

m being a positive integer, and consider the variation of the argument of $mz\, f(z)$ along the circumference $|z| = 1 - \epsilon$. We have, denoting this circumference by Γ,

$$\Delta_\Gamma \operatorname{Arg}[mz\, f(z)] = 2\pi[N + 1 - 1] = 2\pi N.$$

If now we choose m such that $m(1 - \eta)\delta > 2$, we have for $|z| = 1 - \epsilon$,

$$|mz\, f(z)| > m(1 - \eta)\delta > 2,$$

and thus we have also

$$\Delta_\Gamma \operatorname{Arg}[1 + mz\, f(z)] = 2\pi N.$$

But $mz\, f(z) + 1$ has one pole in $|z| < 1 - \epsilon$; hence it has $N + 1$ zeros. Since ϵ can be taken arbitrarily small, it follows that $g(z)$ has $N + 1$ poles for $|z| < 1$. But the expansion of $g(z)$ in the neighborhood of the origin,

$$\sum_0^\infty c_n z^n,$$

has rational integral coefficients. And, in the neighborhood of the circumference $|z| = 1$, $g(z)$ is bounded, since

$$|1 + mzf| > |mzf| - 1 > m(1 - \eta)\delta - 1 > 1.$$

Hence, by Theorem A″ g is a rational function, and so is $f(z)$.

If now $\alpha \neq 0$, let $\alpha = \lambda + \mu i$; we can obviously suppose λ and μ rational, and thus

$$\alpha = \frac{p + qi}{r},$$

p, q, and r being rational integers. Then

$$|rf - (p + qi)| \geq r\delta,$$

and we consider $f^* = rf - (p + qi)$. Then we apply Theorem A″ in the case of Gaussian integers (integers of $K(i)$).

Extensions. The theorem can be extended [13] (1) to the case of the a_n being integers of an imaginary quadratic field, (2) to the case where the number of poles in $|z| < 1$ is infinite (with limit points on $|z| = 1$), (3) to the case of the a_n being integers after only a certain rank $n \geq n_0$, (4) to the case when $z = 0$ is itself a pole. The proof with these extensions does not bring any new difficulties or significant changes into the arguments.

A particular case of the theorem can be stated in the following simple way.

Let

$$f(z) = \sum_0^\infty a_n z^n$$

be a power series with rational integral coefficients, converging for $|z| < 1$. *Let S be the set of values taken by* $f(z)$ *when* $|z| < 1$. *If the derived set S′ is not the whole plane,* $f(z)$ *is a rational function.*

In other words if $f(z)$ is not a rational function, it takes in the unit circle values arbitrarily close to any given number α.

It is interesting to observe that the result would become false if we replace the *whole* unit circle by a circular sector. We shall, in fact, construct a power series with integral coefficients, converging for $|z| < 1$, which is not a rational function, and which is bounded in a certain circular sector of $|z| < 1$. Consider the series

$$f(z) = \sum_{p=0}^\infty \frac{z^{p^2}}{(1 - z)^p}.$$

It converges uniformly for $|z| < r$ if r is any number less than 1. In fact

$$\left|\frac{z^{p^2}}{(1 - z)^p}\right| \leq \frac{r^{p^2}}{(1 - r)^p},$$

which is the general term of a positive convergent series. Hence, $f(z)$ is analytic and regular for $|z| < 1$. It is obvious that its expansion in the unit circle has integral rational coefficients. The function $f(z)$ cannot be rational, for $z = 1$ cannot be a pole of $f(z)$, since $(1 - z)^k f(z)$ increases infinitely as $z \to 1 - 0$ on the real axis, no matter how large the integer k. Finally, $f(z)$ is bounded, say, in the half circle

$$|z| < 1, \quad \Re(z) \leq 0.$$

For, if $\frac{3}{4} < |z| < 1$, say, then

$$|1 - z| > (1 + \tfrac{9}{16})^{\frac{1}{2}} = \tfrac{5}{4},$$

and thus

$$|f(z)| < \sum_0^\infty (\tfrac{4}{5})^p.$$

The function $f(z)$ is even continuous on the arc $|z| = 1$, $\Re(z) \leq 0$.

2. Schlicht power series with integral coefficients [13]

THEOREM II. *Let $f(z)$ be analytic and schlicht (simple) inside the unit circle $|z| < 1$. Let its expansion in the neighborhood of the origin be*

$$f(z) = a_{-1}z^{-1} + \sum_0^\infty a_n z^n.$$

If an integer p exists such that for all $n \geq p$ the coefficients a_n are rational integers (or integers of an imaginary quadratic field), then $f(z)$ is a rational function.

PROOF. Suppose first that $a_{-1} \neq 0$. Then the origin is a pole, and since there can be no other pole for $|z| < 1$, the expansion written above is valid in all the open disc $|z| < 1$. Moreover, the point at infinity being an interior point for the transformed domain, $f(z)$ is bounded for, say, $\frac{1}{2} < |z| < 1$. Hence the power series

$$\sum_0^\infty a_n z^n$$

is bounded in the unit circle, and the nature of its coefficients shows that it is a polynomial, which proves the theorem in this case.

Suppose now that $a_{-1} = 0$. Then $f(z)$ may or may not have a pole inside the unit circle. The point $f(0) = a_0$ is an interior point for the transformed domain. Let $u = f(z)$. To the circle C, $|u - a_0| < \delta$, in the u-plane there corresponds, for δ small enough, a domain D in the z-plane, including the origin, and completely interior, say, to the circle $|z| < \frac{1}{2}$. Now, by Theorem I, if $f(z)$ is not rational, there exists in the ring $\frac{2}{3} < |z| < 1$ a point z_1 such that $|f(z_1) - a_0| < \delta/2$. Then $u_1 = f(z_1)$ belongs to the circle C and consequently there exists in the domain D a point z_2, necessarily distinct from z_1, such that $f(z_2) = u_1 = f(z_1)$. This contradicts the hypothesis that $f(z)$ is schlicht. Hence, $f(z)$ is a rational function.

3. A class of power series with integral coefficients [13]; the class T of algebraic integers and their characterization

Let $f(z)$ be a power series with rational integral coefficients, converging for $|z| < 1$ and admitting at least *one* "exceptional value" in the sense of Theorem I; i.e., we assume that $|f(z) - \alpha| > \delta > 0$ uniformly as $|z| \to 1$. Then $f(z)$ is rational and it is easy to find its form. For

$$f(z) = \frac{P(z)}{Q(z)},$$

P and Q being polynomials with rational integral coefficients, and by Fatou's lemma (see Chapter I) $Q(0) = 1$. The polynomial $Q(z)$ must have no zeros inside the unit circle (P/Q being irreducible) and, since $Q(0) = 1$, it means that all zeros are *on* the unit circle. By a well-known theorem of Kronecker [9] these zeros are all roots of unity unless $Q(z)$ is the constant 1.

Now, suppose that the expansion

$$\sum_0^\infty a_n z^n,$$

with rational integral coefficients, of $f(z)$ is valid only in the neighborhood of the origin, but that $f(z)$ has a simple pole $1/\tau$ ($|\tau| > 1$) and no other singularity for $|z| < 1$.

Suppose again that there exists at least one exceptional value α such that $|f(z) - \alpha| > \delta > 0$ uniformly as $|z| \to 1$. Then $f(z)$ is rational; i.e.,

$$f(z) = \frac{P}{Q},$$

P, Q being polynomials with rational integral coefficients, P/Q irreducible, and $Q(0) = 1$. The point $1/\tau$ is a simple zero for $Q(z)$ and there are no other zeros of modulus less than 1. If $f(z)$ is bounded on the circumference $|z| = 1$, $Q(z)$ has no zeros of modulus 1, all the conjugates of $1/\tau$ lie outside the unit circle, and τ belongs to the class S.

If, on the contrary, $f(z)$ is unbounded on $|z| = 1$, $Q(z)$ has zeros of modulus 1. If all these zeros are roots of unity, $Q(z)$ is divisible by a cyclotomic polynomial, and again τ belongs to the class S. If not, τ is an algebraic integer whose conjugates lie all *inside or on* the unit circle.

We propose to discuss certain properties of this new class of algebraic integers.

DEFINITION. *A number* τ belongs to the class T *if it is an algebraic integer whose conjugates all lie inside or on the unit circle, assuming that some conjugates lie actually on the unit circle (for otherwise* τ *would belong to the class* S*).*

Let $P(z) = 0$ be the irreducible equation determining τ. Since there must be at least one root of modulus 1, and since this root is not ± 1, there must be *two* roots, imaginary conjugates, α and $1/\alpha$ on the unit circle. Since $P(\alpha) = 0$ and $P(1/\alpha) = 0$ and P is irreducible, P is a reciprocal polynomial; τ is its only root outside, and $1/\tau$ its only root inside, the unit circle; τ is real (we may always suppose $\tau > 0$; hence $\tau > 1$). There is an even number of imaginary roots of modulus 1, and the degree of P is even, at least equal to 4. Finally, τ is a unit. If $P(z)$ is of degree $2k$ and if we write

$$y = z + \frac{1}{z},$$

the equation $P(z) = 0$ is transformed into an equation of degree k, $R(y) = 0$, whose roots are algebraic integers, all real. One of these, namely $\tau + \tau^{-1}$, is larger than 2, and all others lie between -2 and $+2$.

We know that the characteristic property of the numbers θ of the class S is that to each $\theta \in S$ we can associate a real $\lambda \neq 0$ such that $\sum \| \lambda\theta^n \|^2 < \infty$; i.e., the series $\sum \| \lambda\theta^n \| z^n$ belongs to the class H^2.†

† Of course, if $\theta \in S$, the series is even *bounded* in $|z| < 1$. But it is *enough* that it should belong to H^2 in order that θ should belong to S.

The corresponding theorem for the class T is the following one.

THEOREM III. *Let τ be a real number > 1. A necessary and sufficient condition for the existence of a real $\mu \neq 0$ such that the power series* †

$$\sum_0^\infty \{\mu\tau^n\} z^n$$

should have its real part bounded above (without belonging to the class H^2) for $|z| < 1$ is that τ should belong to the class T. Then μ is algebraic and belongs to the field of τ.

PROOF. *The condition is necessary.* Let a_n be the integer nearest to $\mu\tau^n$, so that $\mu\tau^n = a_n + \{\mu\tau^n\}$. We have

$$\frac{\mu}{1-\tau z} = \sum_0^\infty a_n z^n + \sum_0^\infty \{\mu\tau^n\} z^n.$$

Now if

$$\frac{\tau+1}{2\tau} < |z| < 1$$

we have

$$|1 - \tau z| > \tfrac{1}{2}(\tau - 1).$$

Hence,

$$\left|\sum_0^\infty a_n z^n + \sum_0^\infty \{\mu\tau^n\} z^n\right| \leq \frac{2|\mu|}{\tau - 1}.$$

Therefore, the real part of

$$f(z) = \sum_0^\infty a_n z^n$$

is bounded below in the ring

$$\frac{\tau+1}{2\tau} < |z| < 1.$$

Since this power series has rational integral coefficients and is regular in $|z| < 1$ except for the pole $1/\tau$, it follows, by Theorem I, that it represents a rational function and, hence, that τ is a number, either of the class S or of the class T. Since $f(z)$ is not in H^2, τ is not in S, and thus belongs to T. The calculation of residues shows that μ is algebraic and belongs to the field of τ.

The condition is sufficient. Let τ be a number of the class T and let $2k$ be its degree. Let

$$\tau^{-1}, \alpha_j, \alpha_j^* \quad (j = 1, 2, \ldots, k-1;\ \alpha_j\alpha_j^* = 1)$$

be its conjugates. Let

$$\sigma = \tau + \tau^{-1}, \quad \rho_j = \alpha_j + \alpha_j^*,$$

so that $\sigma, \rho_1, \rho_2, \ldots, \rho_{k-1}$ are conjugate algebraic integers of degree k.

† See the Introduction (page 1) for the notation $\{a\}$. We recall that $\|a\| = |\{a\}|$.

The determinant

$$\Delta = \begin{vmatrix} 1 & \sigma & \dots & \sigma^{k-1} \\ 1 & \rho_1 & \dots & \rho_1^{k-1} \\ \dots & \dots & \dots & \dots \\ 1 & \rho_{k-1} & \dots & \rho_{k-1}^{k-1} \end{vmatrix}$$

being not zero, we can, by Minkowski's theorem (as given at the beginning of Chapter VI and Appendix, 9), find rational integers $A_1, \dots, A_k$, such that the number

$$\theta = A_1\sigma^{k-1} + \cdots + A_{k-1}\sigma + A_k$$

has its conjugates $\beta_1, \dots, \beta_{k-1}$ all less than 1 in absolute value. In other words, θ is a number of the class S belonging to the field of σ. Its conjugates are all real. Take now

$$\mu = \theta^{2h} \quad \text{and} \quad \gamma_j = \beta_j^{2h},$$

h being a positive integer such that

$$\gamma_1 + \gamma_2 + \cdots + \gamma_{k-1} < \tfrac{1}{8}.$$

Since $\sigma = \tau + \tau^{-1}$ and τ is a unit, μ is an algebraic integer of the field of τ, $K(\tau)$, and the numbers

$$\mu \text{ itself}, \gamma_1, \gamma_1, \gamma_2, \gamma_2, \dots, \gamma_{k-1}, \gamma_{k-1}$$

correspond to μ in the conjugate fields

$$K(\tau^{-1}), K(\alpha_1), K(\alpha_1^*), \dots, K(\alpha_{k-1}), K(\alpha_{k-1}^*)$$

respectively. It follows that the function

$$f(z) = \frac{\mu}{1 - \tau z} + \frac{\mu}{1 - \tau^{-1}z} + \sum_1^{k-1} \frac{\gamma_j}{1 - \alpha_j z} + \sum_1^{k-1} \frac{\gamma_j}{1 - \alpha_j^* z}$$

has, in the neighborhood of the origin, an expansion

$$\sum_0^\infty a_n z^n$$

with rational integral coefficients. The only singularity of $f(z)$ for $|z| < 1$ is the pole $1/\tau$. We have

$$\sum_0^\infty (a_n - \mu\tau^n)z^n = \sum_0^\infty a_n z^n - \frac{\mu}{1 - \tau z} = \frac{\mu}{1 - \tau^{-1}z} + \sum_1^{k-1} \frac{\gamma_j}{1 - \alpha_j z} + \sum_1^{k-1} \frac{\gamma_j}{1 - \alpha_j^* z}.$$

By well-known properties of linear functions we have for $|\alpha| = 1$ and $|z| < 1$

$$\Re\left(\frac{1}{1 - \alpha z}\right) \geq \frac{1}{2}$$

and

$$\Re\left(\frac{1}{1 - \tau^{-1}z}\right) \geq \frac{\tau}{\tau + 1}.$$

Therefore, since $\gamma_j > 0$, we have for $|z| < 1$

$$\Re\left\{\sum_0^\infty (a_n - \mu\tau^n)z^n\right\} \geq \frac{\mu\tau}{\tau+1} + \sum_1^{k-1} \gamma_j > \frac{\mu\tau}{\tau+1}.$$

On the other hand,

$$a_n = \mu\tau^n + \mu\tau^{-n} + \sum_1^{k-1} \gamma_j(\alpha_j^n + \alpha_j^{*n})$$

and, since $|\alpha_j^n + \alpha_j^{*n}| < 2$,

$$\left|\sum_1^{k-1} \gamma_j(\alpha_j^n + \alpha_j^{*n})\right| < 2\sum_1^{k-1} \gamma_j < \tfrac{1}{4}.$$

Take now for m the smallest integer such that

$$\frac{\mu}{\tau^m} < \frac{1}{4}; \quad \text{i.e., } m = \left[\frac{\log 4\mu}{\log \tau}\right] + 1.$$

Then, for $n \geq m$

$$|a_n - \mu\tau^n| < \tfrac{1}{2}; \quad \text{i.e., } a_n - \mu\tau^n = -\{\mu\tau^n\}.$$

Therefore, we can write

$$\Re\left\{\sum_0^{m-1} (a_n - \mu\tau^n)z^n - \sum_m^\infty \{\mu\tau^n\}\, z^n\right\} > \frac{\mu\tau}{\tau+1}.$$

On the other hand, since for all n

$$|a_n - \mu\tau^n| < \frac{\mu}{\tau^n} + \frac{1}{4},$$

we have for $|z| < 1$

$$\left|\sum_0^{m-1} (a_n - \mu\tau^n)z^n\right| < \frac{m}{4} + \mu\sum_0^{m-1} \frac{1}{\tau^n} < \frac{m}{4} + \frac{\mu\tau}{\tau-1},$$

whence, finally,

$$\Re\left\{\sum_m^\infty \{\mu\tau^n\}\, z^n\right\} < \frac{m}{4} + \frac{\mu\tau}{\tau-1} - \frac{\mu\tau}{\tau+1}$$

$$= \frac{m}{4} + \frac{2\mu\tau}{\tau^2-1}.$$

Thus

$$\Re\left\{\sum_0^\infty \{\mu\tau^n\}\, z^n\right\} < \frac{3m}{4} + \frac{2\mu\tau}{\tau^2-1} = A(\mu, \tau),$$

where A is a function of μ and τ only, which proves the second part of our theorem.

4. Properties of the numbers of the class T

THEOREM IV. *Every number of the class S*† *is a limit point of numbers of the class T on both sides* [13].

PROOF. Let θ be a number of the class S, root of the irreducible polynomial

$$P(z) = z^p + c_1 z^{p-1} + \cdots + c_p$$

with rational integral coefficients. Let $Q(z)$ be the reciprocal polynomial.

We suppose first that θ is not a quadratic unit, so that Q and P will not be identical.

We denote by m a positive integer, and let

$$R_m(z) = z^m P(z) + Q(z).$$

Then $R_m(z)$ is a reciprocal polynomial whose zeros are algebraic integers.

We denote by ϵ a positive number and consider the equation

$$(1+\epsilon)z^m P + Q = 0.$$

Since for $|z| = 1$ we have $|P| = |Q|$, it follows by Rouché's theorem that *in* the circle $|z| = 1$ the number of roots of the last equation is equal to the number of roots of $z^m P$, that is to say, $m + p - 1$. As $\epsilon \to 0$, these roots vary continuously. Hence, for $\epsilon = 0$ we have $m + p - 1$ roots with modulus ≤ 1 and, hence, at most one root *outside* the unit circle.‡

It is easy to show now that the root of $R_m(z)$ with modulus *larger* than 1 actually exists. In fact, we have first

$$R_m(\theta) = Q(\theta) \neq 0,$$

since θ is not quadratic. On the other hand, it is easily seen that $P'(\theta) > 0$. We fix $\sigma > 0$ small enough for $P'(z)$ to have no zeros on the real axis in the interval

$$\theta - \sigma \leq z \leq \theta + \sigma.$$

We suppose that in this interval $P'(z) > \mu$, μ being a positive number fixed as soon as σ is fixed.

If we take δ real and $|\delta| < \sigma$, $P(\theta + \delta)$ has the sign of δ and is in absolute value not less than $|\delta|\mu$. Hence, taking e.g.,

$$|\delta| = \frac{1}{\sqrt{m}},$$

† We recall that we do not consider the number 1 as belonging to the class S (see Chapter I).

‡ This proof, much shorter and simpler than the original one, has been communicated to me by Prof. Hirschman, during one of my lectures at the Sorbonne.

we see that for m large enough

$$R_m(\theta+\delta) = (\theta+\delta)^m P(\theta+\delta) + Q(\theta+\delta)$$

has the sign of δ. Taking $\delta Q(\theta) < 0$, we see that $R_m(\theta)$ and $R_m(\theta+\delta)$ are, for m large enough, of opposite sign, so that $R_m(z)$ has a root τ_m

$$\text{between } \theta \text{ and } \theta + m^{-\frac{1}{2}} \text{ if } Q(\theta) < 0,$$

and

$$\text{between } \theta - m^{-\frac{1}{2}} \text{ and } \theta \text{ if } Q(\theta) > 0.$$

Hence, $\tau_m \to \theta$ as $m \to \infty$.

This proves, incidentally, since we can have a sequence of τ_m all different tending to θ, that there exist numbers of the class T of arbitrarily large degree. It proves also that τ_m has, actually, conjugates of modulus 1, for m large enough, for evidently τ_m cannot be constantly quadratic (see Appendix, 5).

To complete the proof for θ not quadratic, we consider, instead of $z^m P + Q$, the polynomial

$$\frac{z^m P - Q}{z - 1},$$

which is also reciprocal, and we find a sequence of numbers of the class T approaching θ from the other side.

Suppose now that θ is a quadratic unit. Thus θ is a quadratic integer > 1, with conjugate $\frac{1}{\theta}\cdot$ Then $\theta + \theta^{-1}$ is a rational integer $r \geq 3$. Denote by $T_m(x)$ the first Tchebycheff polynomial of degree m (i.e., $T_m(x) = 2\cos m\varphi$ for $x = 2\cos(\varphi)$). T_m has m distinct real zeros between -2 and $+2$. The equation

$$(x - r)\,T_m(x) - 1 = 0$$

has then $m - 1$ real roots (algebraic integers) between -2 and $+2$, and one real root between r and $r + \epsilon_m$ ($\epsilon_m > 0$, $\epsilon_m \to 0$ as $m \to \infty$). Putting

$$x = y + \frac{1}{y},$$

we get an equation in y which gives us a number of the class T approaching θ from the right as $m \to \infty$.

We get numbers of T approaching θ from the left if we start from the equation

$$(x - r)T_m(x) + 1 = 0.$$

This completes the proof of the theorem.

We do not know whether numbers of T have limit points other than numbers of S.

5. Arithmetical properties of the numbers of the class T

We have seen at the beginning of Chapter I that, far from being uniformly distributed, the powers θ^n of a number θ of the class S tend to zero modulo 1.

On the contrary, the powers τ^m of a number τ of the class T are, modulo 1, everywhere dense in the interval (0, 1). In order to prove this, let us consider a number $\tau > 1$ of the class T, root of an irreducible equation of degree $2k$. We denote the roots of this equation by

$$\tau, \quad \frac{1}{\tau}, \quad \alpha_1, \quad \alpha_2, \quad \ldots, \quad \alpha_{k-1}, \qquad \bar{\alpha}_1, \quad \bar{\alpha}_2, \quad \ldots, \quad \bar{\alpha}_{k-1},$$

where $|\alpha_j| = 1$ and $\bar{\alpha}_j = \alpha_j^{-1}$ is the imaginary conjugate of α_j. We write

$$\alpha_j = e^{2\pi i\omega_j}.$$

Our first step will be to show that the ω_j $(j = 1, 2, \ldots, k-1)$ and 1 are linearly independent.† For suppose, on the contrary, the existence of a relation

$$A_0 + A_1\omega_1 + \cdots + A_{k-1}\omega_{k-1} = 0,$$

the A_j being rational integers. Then

$$e^{2\pi i(A_0 + \cdots + A_{k-1}\omega_{k-1})} = 1$$

or

$$(1) \qquad \alpha_1^{A_1}\alpha_2^{A_2}\cdots\alpha_{k-1}^{A_{k-1}} = 1.$$

Since the equation considered is irreducible, it is known ([1] and Appendix, 6) that its Galois group is transitive; i.e., there exists an automorphism σ of the Galois group sending, e.g., the root α_1 into the root τ. This automorphism can not send any α_j into $1/\tau$; for, since $\sigma(\alpha_1) = \tau$,

$$\sigma\left(\frac{1}{\alpha_1}\right) = \frac{1}{\tau},$$

and thus this would imply

$$\alpha_j = \frac{1}{\alpha_1},$$

which is not the case. Thus the automorphism applied to (1) gives

$$\tau^{A_1}\alpha_2'^{A_2}\cdots\alpha_{k-1}'^{A_{k-1}} = 1$$

if $\sigma(\alpha_j) = \alpha_j'$ $(j \neq 1)$. This is clearly impossible since $\tau > 1$ and $|\alpha_s'| = 1$. Hence, we have proved the linear independence of the ω_j and 1.

Now, we have, modulo 1,

$$\tau^m + \frac{1}{\tau^m} + \sum_{j=1}^{k-1}(e^{2\pi im\omega_j} + e^{-2\pi im\omega_j}) \equiv 0$$

† This argument is due to Pisot.

or

$$\tau^m + 2\sum_{j=1}^{k-1} \cos 2\pi m\omega_j \to 0 \pmod 1$$

as $m \to \infty$. But by the well-known theorem of Kronecker on linearly independent numbers ([2] and Appendix, 8) we can determine the integer m, arbitrarily large, such that

$$2\sum_{j=1}^{k-1} \cos 2\pi m\omega_j$$

will be arbitrarily close to any number given in advance (mod 1). It is enough to take m, according to Kronecker, such that

$$|m\omega_1 - \delta| < \epsilon \pmod 1$$

$$|m\omega_j - \tfrac{1}{4}| < \epsilon \pmod 1 \quad (j = 2, 3, \ldots, k - 1).$$

We have thus proved that the $\{\tau^m\}$ (mod 1) are everywhere dense.

The same argument applied to $\lambda\tau^m$, λ being an integer of the field of τ, shows that $\lambda\tau^m$ (mod 1) is everywhere dense in a certain interval.

THEOREM V. *Although the powers τ^m of a number τ of the class T are, modulo* 1, *everywhere dense, they are not uniformly distributed in* (0, 1).

In order to grasp better the argument, we shall first consider a number τ of the class T of the 4th degree. In this case the roots of the equation giving τ are

$$\tau, \quad \frac{1}{\tau}, \quad \alpha, \quad \bar{\alpha} = \frac{1}{\alpha} \quad (|\alpha| = 1),$$

and we have, m being a positive integer,

$$\tau^m + \frac{1}{\tau^m} + \alpha^m + \bar{\alpha}^m \equiv 0 \pmod 1.$$

Writing $\alpha = e^{2\pi i\omega}$, we have

$$\tau^m + \frac{1}{\tau^m} + 2\cos 2\pi m\omega \equiv 0 \pmod 1.$$

The number ω is irrational. This is a particular case of the above result, where we prove linear independence of $\omega_1, \omega_2, \ldots, \omega_{k-1}$, and 1. One can also argue in the following way. If ω were rational, α would be a root of 1, and the equation giving τ would not be irreducible.

Now, in order to prove the nonuniform distribution of τ^m (mod 1), it is enough to prove the nonuniform distribution of $2\cos 2\pi m\omega$. This is a consequence of the more general lemma which follows.

LEMMA. *If the sequence* $\{u_n\}_1^\infty$ *is uniformly distributed modulo* 1, *and if* $\omega(x)$ *is a continuous function, periodic with period* 1, *the sequence* $\omega(u_n) = v_n$ *is uniformly distributed if and only if the distribution function of* $\omega(x)$ (mod 1) *is linear.*†

PROOF of the lemma. Let (a, b) be any subinterval of $(0, 1)$ and let $\chi(x)$ be a periodic function, with period 1, equal for $0 \leq x < 1$ to the characteristic function of (a, b). The uniform distribution modulo 1 of $\{v_n\}$ is equivalent to

$$\lim \frac{\sum_1^N \chi(v_n)}{N} = b - a.$$

But, owing to the uniform distribution of $\{u_n\}$,

$$\lim \frac{1}{N}\sum_1^N \chi(v_n) = \lim \frac{1}{N}\sum_1^N \chi[\omega(u_n)] = \int_0^1 \chi(\omega(x))\, dx.$$

Let $\omega^*(x) \equiv \omega(x) \pmod 1$, $0 \leq \omega^*(x) < 1$. The last integral is

$$\int_0^1 \chi(\omega^*(x))dx = \text{meas } E\,\{a < \omega^*(x) < b\}.$$

Hence,

$$\text{meas } E\,\{a < \omega^*(x) < b\} = b - a, \tag{2}$$

which proves the lemma.

An alternative necessary and sufficient condition for the uniform distribution modulo 1 of $v_n = \omega(u_n)$ is that

$$\int_0^1 e^{2\pi i h\omega(x)}\, dx = 0 \tag{3}$$

for all integers $h \neq 0$.

For the uniform distribution of $\{v_n\}$ is equivalent to

$$\lim \frac{1}{N}\sum_1^N e^{2\pi i h\omega(u_n)} = \lim \frac{1}{N}\sum_1^N e^{2\pi i h v_n} = 0$$

by Weyl's criterion. But

$$\lim \frac{1}{N}\sum_1^N e^{2\pi i h\omega(u_n)} = \int_0^1 e^{2\pi i h\omega(x)}\, dx.$$

Hence, we have the result, and it can be proved directly without difficulty, considering again $\omega^*(x)$, that (3) is equivalent to (2).

In our case $\{u_m\} = \{m\omega\}$ is uniformly distributed modulo 1, and it is enough to remark that the function $2 \cos 2\pi x$ has a distribution function (mod 1) which

† No confusion can arise from the notation $\omega(x)$ for the distribution function and the *number* ω occurring in the proof of the theorem.

is not linear. This can be shown by direct computation or by remarking that

$$\int_0^1 e^{4\pi i h \cos 2\pi x} = J_0(4\pi h)$$

is not zero for all integers $h \neq 0$.

In the general case (τ not quadratic) if $2k$ is the degree of τ, we have, using the preceding notations,

$$\tau^m + \frac{1}{\tau^m} + \sum_{j=1}^{k-1} \alpha_j^m + \sum_{j=1}^{k-1} \alpha_j^{-m} \equiv 0 \pmod 1$$

or

$$\tau^m + \frac{1}{\tau^m} + \sum_{j=1}^{k-1} 2 \cos 2\pi m\omega_j \equiv 0 \pmod 1$$

and we have to prove that the sequence

$$v_m = 2 \cos 2\pi m\omega_1 + \cdots + 2 \cos 2\pi m\omega_{k-1}$$

is not uniformly distributed modulo 1.

We use here a lemma analogous to the preceding one.

LEMMA. *If the p-dimensional vector* $\{u_n^j\}_{n=1}^{\infty}$ $(j = 1, 2, \ldots p)$ *is uniformly distributed modulo* 1 in R^p, *the sequence*

$$v_n = \omega(u_n^1) + \omega(u_n^2) + \cdots + \omega(u_n^p)$$

where $\omega(x)$ *is continuous with period* 1 *is uniformly distributed if and only if condition* (2) *or the equivalent condition* (3) *is satisfied.*

PROOF of the lemma. It is convenient here to use the second form of the proof. The condition is

$$\frac{1}{N} \sum_{1}^{N} e^{2\pi i h v_n} \to 0 \quad (h \text{ is any integer} \neq 0).$$

But

$$\frac{1}{N} \sum_{n=1}^{N} e^{2\pi i h \{\omega(u_n^1) + \cdots + \omega(u_n^p)\}} \to \left\{ \int_0^1 e^{2\pi i h \omega(x)} \, dx \right\}^p .$$

Hence the lemma.

Theorem V about τ^m follows from the fact that $\{m\omega_1, m\omega_2, \ldots, m\omega_{k-1}\}$ is uniformly distributed in the unit torus of R^{k-1} owing to the fact that $\omega_1, \ldots,$ ω_{k-1}, and 1 are linearly independent. This completes the proof.

EXERCISE

Show that any number τ of the class T is the quotient θ/θ' of two numbers of the class S belonging to the field of τ. (For this and other remarks, see [13].)

Chapter IV

A CLASS OF SINGULAR FUNCTIONS; BEHAVIOR OF THEIR FOURIER–STIELTJES TRANSFORMS AT INFINITY

1. Introduction

By a *singular function* $f(x)$ we shall mean, in what follows, a singular continuous monotonic function (e.g., nondecreasing), bounded, and whose derivative vanishes for almost all (in the Lebesgue sense) values of the real variable x.

A wide class of singular functions is obtained by constructing, say, in $(0, 2\pi)$ a perfect set of measure zero, and by considering a nondecreasing continuous function $f(x)$, constant in every interval contiguous to the set (but not everywhere).

A very interesting and simple example of perfect sets to be considered is the case of symmetrical perfect sets with constant ratio of dissection. Let ξ be a positive number, strictly less than $\frac{1}{2}$, and divide the fundamental interval, say, $(0, 2\pi)$, into three parts of lengths proportional to ξ, $1 - 2\xi$, and ξ respectively. Remove the central open interval ("black" interval). Two intervals ("white" intervals) are left on which we perform the same operation. At the kth step we are left with 2^k white intervals, each one of length $2\pi\xi^k$. Denote by E_k the set of points belonging to these 2^k closed white intervals. Their left-hand end points are given by the formula

$$x = 2\pi[\epsilon_1(1-\xi) + \epsilon_2\xi(1-\xi) + \cdots + \epsilon_k\xi^{k-1}(1-\xi)], \tag{1}$$

where the ϵ_i are 0 or 1. The intersection of all E_k is a perfect set E of measure equal to

$$2\pi \lim_{k=\infty} (\xi^k 2^k) = 0$$

and whose points are given by the infinite series

$$x = 2\pi[\epsilon_1(1-\xi) + \epsilon_2\xi(1-\xi) + \cdots + \epsilon_k\xi^{k-1}(1-\xi) + \cdots], \tag{2}$$

where the ϵ_i can take the values 0 or 1. The reader will recognize that the classical Cantor's ternary set is obtained by taking $\xi = \frac{1}{3}$.

We define now, when $x \in E$, a function $f(x)$ given by

$$f(x) = \frac{\epsilon_1}{2} + \frac{\epsilon_2}{2^2} + \cdots + \frac{\epsilon_k}{2^k} + \cdots,$$

when x is given by (2). It is easily seen that at the end points of a black interval (e.g., $\epsilon_1 = 0$, $\epsilon_2 = \epsilon_3 = \cdots = 1$ and $\epsilon_1 = 1$, $\epsilon_2 = \epsilon_3 = \cdots = 0$) $f(x)$ takes the same value. We then define $f(x)$ in this interval as a constant equal to this common value. The function $f(x)$ is now defined for $0 \leq x \leq 2\pi$ ($f(0) = 0$, $f(2\pi) = 1$), is continuous, nondecreasing, and obviously singular. We shall call it the "Lebesgue function" associated with the set E.

The Fourier-Stieltjes coefficients of df are given by

$$c_n = (2\pi)^{-1} \int_0^{2\pi} e^{nix}\, df(x), \tag{3}$$

and, likewise, the Fourier-Stieltjes transform of df is defined by

$$\begin{aligned}\gamma(u) &= (2\pi)^{-1} \int_{-\infty}^{\infty} e^{uix}\, df(x)\\ &= (2\pi)^{-1} \int_0^{2\pi} e^{uix}\, df(x)\end{aligned}$$

for the continuous parameter u, f being defined to be equal to 0 in $(-\infty, 0)$ and to 1 in $(2\pi, \infty)$.

One can easily calculate the Riemann-Stieltjes integral in (3) by remarking that in each "white" interval of the kth step of the dissection f increases by $1/2^k$. The origins of the intervals are given by (1), or, for the sake of brevity, by

$$x = 2\pi[\epsilon_1 r_1 + \cdots + \epsilon_k r_k],$$

with $r_k = \xi^{k-1}(1-\xi)$. Hence an approximate expression of the integral

$$\int_0^{2\pi} e^{nix}\, df$$

is

$$\frac{1}{2^k} \sum e^{2\pi ni(\epsilon_1 r_1 + \cdots + \epsilon_k r_k)},$$

the summation being extended to the 2^k combinations of $\epsilon_j = 0, 1$. This sum equals

$$\frac{1}{2^k} \prod_{s=1}^{k} (1 + e^{2\pi nir_s}) = e^{\pi ni \sum_1^k r_s} \prod_{s=1}^{k} \cos \pi nr_s.$$

Since $\sum_1^\infty r_s = 1$, we have

$$e^{-\pi ni}\, 2\pi c_n = \prod_{k=1}^{\infty} \cos \pi nr_k = \prod_{k=1}^{\infty} \cos \pi n\xi^{k-1}(1-\xi) \tag{4}$$

and likewise

$$e^{-\pi ui}\, 2\pi\gamma(u) = \prod_{k=1}^{\infty} \cos \pi u\xi^{k-1}(1-\xi). \tag{5}$$

2. The problem of the behavior at infinity

It is well known in the elementary theory of trigonometric series that if f is absolutely continuous, the Fourier-Stieltjes transform

$$\gamma(u) = (2\pi)^{-1} \int_0^{2\pi} e^{uix}\, df$$

tends to zero as $|u| \to \infty$, because in this case $\gamma(u)$ is nothing but the ordinary Fourier transform of a function of the class L. The situation is quite different if f is continuous, but singular. In this case $\gamma(u)$ need not tend to zero, although there do exist singular functions for which $\gamma(u) \to 0$ ([17], and other examples in this chapter). The same remarks apply to the Fourier-Stieltjes coefficients c_n.

The problem which we shall solve here is the following one. Given a symmetrical perfect set with constant ratio of dissection ξ, which we shall denote by $E(\xi)$, we construct the Lebesgue function f connected with it, and we try to determine for what values of ξ the Fourier-Stieltjes transform (5) (or the Fourier-Stieltjes coefficient (4)) tends or does not tend to zero as $|u|$ (or $|n|$) increases infinitely.

We shall prove first the following general theorem.

THEOREM I. *For any function of bounded variation f the Riemann-Stieltjes integrals*

$$2\pi c_n = \int_0^{2\pi} e^{nix}\, df \quad \textit{and} \quad 2\pi\gamma(u) = \int_0^{2\pi} e^{uix}\, df$$

tend or do not tend to zero together when $|n|$ or $|u|$ tends to ∞.

Since it is obvious that $\gamma(u) = o(1)$ implies $c_n = o(1)$, we have only to prove the converse proposition. We shall base this proof on the following lemma, interesting in itself.

LEMMA. *Let $f(x)$ be a function of bounded variation such that, as $|n| \to \infty$,*

$$\int_0^{2\pi} e^{nix}\, df \to 0. \tag{6}$$

Let $B(x)$ be any function such that the Lebesgue-Stieltjes integral

$$\int_0^{2\pi} B(x)\,df(x)$$

has a meaning. Then the integral

$$\int_0^{2\pi} e^{nix} B(x)\,df$$

tends also to zero for $|n| = \infty$.

PROOF of the lemma. We observe first that by the properties of the Lebesgue-Stieltjes integral, there exists a step function $T(x)$ such that

$$\int_0^{2\pi} | B(x) - T(x) | \, df < \epsilon, \tag{7}$$

ϵ being arbitrarily small. Secondly, by a well-known theorem of Wiener [17], the condition (6) implies that f is continuous. Hence, in (7) we can replace $T(x)$ by a trigonometric polynomial $P(x)$. But (6) implies

$$\int_0^{2\pi} e^{nix} P(x) df \to 0.$$

Hence, ϵ being arbitrarily small in (7), we have

$$\int_0^{2\pi} B(x) e^{nix} \, df \to 0$$

as stated in the lemma.

PROOF of Theorem I. Suppose that

$$c_n \to 0 \text{ as } | n | \to \infty.$$

If $\gamma(u)$ does not tend to zero as $| u | \to \infty$, we can find a sequence

$$\{u_k\}_{k=1}^{\infty} \text{ with } | u_k | \to \infty$$

such that

$$\left| \int_0^{2\pi} e^{u_k ix} \, df \right| \geq \delta > 0.$$

Let

$$u_k = n_k + \alpha_k,$$

n_k being an integer and $0 \leq \alpha_k < 1$. By extracting, if necessary, a subsequence from $\{u_k\}$, we can suppose that α_k tends to a limit α. We would then have

$$\left| \int_0^{2\pi} e^{n_k ix} e^{\alpha ix} \, df \right| \geq \frac{\delta}{2} > 0,$$

which is contrary to the lemma, since $c_n \to 0$ and $e^{\alpha ix}$ is continuous.

It follows now that in order to study the behavior of c_n or $\gamma(u)$, it is enough to study

$$\Gamma(u) = \prod_{k=1}^{\infty} \cos \pi u \xi^k \tag{8}$$

when $u \to \infty$.

THEOREM II. *The infinite product* $\Gamma(u)$ *tends to zero as* $u \to \infty$ *if and only if* $1/\xi$ *is not a number of the class S (as defined in Chapter I). We suppose here* $\xi \neq \frac{1}{2}$.

Remark. We have seen that the expressions (4) and (5) represent respectively the Fourier-Stieltjes coefficient and the Fourier-Stieltjes transform of the Lebesgue function constructed on the set $E(\xi)$ if $0 < \xi < \frac{1}{2}$. Nevertheless, it is easy to see that in order that the infinite products (4), (5), (8) have a meaning, it is enough to suppose that $0 < \xi < 1$. For example, $\Gamma(u)$ still represents a Fourier-Stieltjes transform if only $0 < \xi < 1$, namely the transform of the monotonic function which is the convolution of an infinity of discontinuous measures (mass $\frac{1}{2}$ at each of the two points $\pi\xi^k, -\pi\xi^k$).

Our theorem being true in the general case $0 < \xi < 1$, we shall only assume this condition to prove it.

PROOF of Theorem II. If $\Gamma(u) \neq o(1)$ for $u = \infty$, we can find an infinite increasing sequence of numbers u_s such that

$$|\Gamma(u_s)| \geq \delta > 0.$$

Writing $1/\xi = \theta$ $(\theta > 1)$, we can write

$$u_s = \lambda_s \theta^{m_s},$$

where the m_s are natural integers increasing to ∞, and $1 \leq \lambda_s < \theta$.

By extracting, if necessary, a subsequence from $\{u_s\}$, we can suppose that $\lambda_s \to \lambda$ $(1 \leq \lambda \leq \theta)$. We write

$$|\Gamma(u_s)| \leq \cos \pi\lambda_s \cos \pi\lambda_s\theta \cdots \cos \pi\lambda_s\theta^{m_s},$$

whence

$$\prod_{q=0}^{m_s} [1 - \sin^2 \pi\lambda_s\theta^q] \geq \delta^2,$$

and, since $1 + x < e^x$,

$$e^{-\sum_{q=0}^{m_s} \sin^2 \pi\lambda_s\theta^q} \geq \delta^2;$$

that is to say,

$$\sum_{q=0}^{m_s} \sin^2 \pi\lambda_s\theta^q \leq \log(1/\delta^2).$$

Choosing any $r > s$, we have

$$\sum_{q=0}^{m_s} \sin^2 \pi\lambda_r\theta^q \leq \sum_{q=0}^{m_r} \sin^2 \pi\lambda_r\theta^q \leq \log(1/\delta^2).$$

Keeping now s fixed and letting $r \to \infty$, we have

$$\sum_{q=0}^{m_s} \sin^2 \pi\lambda\theta^q \leq \log(1/\delta^2),$$

and, since s is arbitrarily large,

$$\sum_{q=0}^{\infty} \sin^2 \pi\lambda\theta^q \leq \log(1/\delta^2),$$

which, according to the results of Chapter I, shows that $\theta = \xi^{-1}$ belongs to the class S.

We have thus shown that $\Gamma(u) \neq o(1)$ implies that $\theta \in S$.

Conversely, if $\theta \in S$ and $\theta \neq 2$, then $\Gamma(u)$ does not tend to zero. (Remark that if $\xi = \frac{1}{2}$, the Fourier-Stieltjes coefficient c_n of (4) is zero for all $n \neq 0$ and then $f(x) = x$ $(0 \leq x \leq 2\pi)$.)

Supposing now $\theta = \xi^{-1} \neq 2$, we have

$$\Gamma(\theta^k) = |\cos \pi\theta \cos \pi\theta^2 \cdots \cos \pi\theta^k| \cdot \left|\cos\frac{\pi}{\theta} \cos\frac{\pi}{\theta^2} \cdots \cos\frac{\pi}{\theta^k} \cdots\right| \cdot$$

Since $\theta \in S$, we have $\sum \sin^2 \pi\theta^n < \infty$. Hence, the infinite product

$$\prod_{m=1}^{\infty} \cos^2 \pi\theta^m$$

converges to a number $A \neq 0$ (except if $\theta^q = h + \frac{1}{2}$, h being a natural integer, but this is incompatible with the fact that $\theta \in S$). Hence,

$$|\Gamma(\theta^k)| \geq \sqrt{A} \left|\cos\frac{\pi}{\theta} \cos\frac{\pi}{\theta^2} \cdots\right|$$

and the last product converges to a number $B > 0$, since $\theta \neq 2$ ($\theta^q = 2$ is impossible for $q > 1$ if $\theta \in S$). Hence,

$$|\Gamma(\theta^k)| \geq B\sqrt{A},$$

which completes the proof of Theorem II.

Chapter V

THE UNIQUENESS OF THE EXPANSION IN TRIGONOMETRIC SERIES; GENERAL PRINCIPLES

1. Fundamental definitions and results

Let us consider a trigonometric series

$$(S) \qquad \sum_{0}^{\infty} (a_n \cos nx + b_n \sin nx),$$

where the variable x is real. The classical theory of Cantor shows [17] that if this series converges everywhere to zero, it vanishes identically.

Cantor himself has generalized this result by proving that if (S) converges to zero for all values of x except for an exceptional set E containing a *finite* number of points x, then the conclusion is the same one, i.e.,

$$a_n = 0, \ b_n = 0 \text{ for all } n.$$

Cantor proved also that the conclusion is still valid if E is infinite, provided that the derived set E' is finite, or even provided that any one of the derived sets of E (of *finite* or *transfinite* order) is empty, in other words if E is a denumerable set which is *reducible* [17].

The results of Cantor go back to the year 1870. Not until 1908 was it proved by W. H. Young that the result of Cantor can be extended to the case where E is *any denumerable set* (even if it is not reducible).

The preceding results lead to the following definition.

DEFINITION. *Let E be a set of points x in $(0, 2\pi)$. Then E is a* set of uniqueness *(set U) if no trigonometric series exists (except vanishing identically) converging to zero everywhere, except, perhaps, for $x \in E$. Otherwise E will be called* set of multiplicity *(set M).*

We have just seen that any denumerable set is a set U. On the other hand, as we shall easily show (page 44):

If E is of positive measure, E is a set M.

It is, therefore, natural to try to characterize the sets of measure zero by classifying them in "sets U" and "sets M." We shall give a partial solution of this problem in the next two chapters, but we must begin here by recalling certain classical theorems of the theory of trigonometric series of Riemann [17].

DEFINITIONS. (*a*) *Given any function* $G(x)$ *of the real variable* x, *we shall write*

$$\frac{1}{h^2}\Delta^2 G(x, h) = \frac{G(x+h) + G(x-h) - 2G(x)}{h^2},$$

and, if this expression tends for a given fixed x *to a limit* λ, *as* $h \to 0$, *we shall say that* $G(x)$ *has, at the point* x, *a* second generalized derivative equal to λ.

(*b*) *If, at a given point* x, *the expression*

$$\frac{1}{h}\Delta^2 G(x, h) = \frac{G(x+h) + G(x-h) - 2G(x)}{h}$$

tends to zero as $h \to 0$, *we shall say that* $G(x)$ *is* smooth *at the point* x.

THEOREM I (Cantor-Lebesgue). *If the trigonometric series*

$$\frac{1}{2}a_0 + \sum_{1}^{\infty} (a_n \cos nx + b_n \sin nx) \tag{1}$$

converges in a set of positive measure, its coefficients a_n *and* b_n *tend to zero.*

DEFINITION. *If we integrate the series* (1) *formally twice, assuming that* $a_n \to 0$, $b_n \to 0$, *we obtain the continuous function*

$$F(x) = \frac{1}{4}a_0x^2 - \sum_{1}^{\infty} \frac{(a_n \cos nx + b_n \sin nx)}{n^2}, \tag{2}$$

the last series being uniformly convergent. If, at a given point x, $F(x)$ *has a second generalized derivative equal to* s, *we shall say that the series* (1) *is* summable-Riemann (*or* summable-R) *and that its* sum is s.

THEOREM II. *If the series* (1) ($a_n, b_n \to 0$) *converges to* s *at the point* x, *it is also summable-R to* s *at this point.*

THEOREM IIA. *If the series* (1) *with coefficients tending to zero is summable-R to zero for all the points of an interval, it converges to zero in this interval* (*consequence of the principle of* "*localization*").

THEOREM III. *The function* $F(x)$ (*always assuming* $a_n \to 0$, $b_n \to 0$) *is smooth at every point* x.

THEOREM IV. *Let* $G(x)$ *be continuous in an interval* (a, b). *If the generalized second derivative exists and is zero in* (a, b), $G(x)$ *is linear in* (a, b).

THEOREM V. *Theorem* IV *remains valid if one supposes that the generalized second derivative exists and is zero except at the points of a denumerable set* E, *provided that at these points* G *is smooth.*

Historically, this last theorem was proved first by Cantor (*a*) when E is finite, (*b*) when E is reducible, i.e., has a derived set of finite or transfinite order which is empty. It was extended much later by Young to the general case where E is supposed only to be denumerable.

From Theorem V we deduce finally:

THEOREM VI. *If the series* (1) *converges to* 0 *at all points of* $(0, 2\pi)$ *except perhaps when x belongs to a denumerable set E, the series vanishes identically.* In other words every denumerable set is a set U, which is the above stated result.

PROOF. This follows immediately as a consequence of Theorems II, III, and V. For the application of these theorems shows that the function $F(x)$ of (2) is linear. Hence, for all x,

$$\sum_{1}^{\infty} \frac{a_n \cos nx + b_n \sin nx}{n^2} = \frac{1}{4} a_0 x^2 - Ax - B$$

and the periodicity of the series implies $a_0 = A = 0$; next, the series being uniformly convergent, $B = 0$ and $a_n = b_n = 0$ for all n.

We shall now prove the theorem on page 42:

THEOREM. *Every set of positive measure is a set M.*

PROOF. Let $E \subset (0, 2\pi)$ and $|E| > 0$. It will be enough to prove that there exists a trigonometric series (not vanishing identically) and converging to zero in the complementary set of E, that is, CE.

Let P be perfect such that $P \subset E$, and $|P| > 0$. Let $\chi(x)$ be its characteristic function. In an interval Δ contiguous to P, one has $\chi(x) = 0$; hence the Fourier series of $\chi(x)$,

$$\frac{\alpha_0}{2} + \sum_{0}^{\infty} (\alpha_n \cos nx + \beta_n \sin nx) \sim \chi(x),$$

converges to zero in Δ. Hence it converges to zero in CP, and also in $CE \subset CP$. But this series does not vanish identically, since

$$\alpha_0 = \frac{1}{\pi} \int_0^{2\pi} \chi(x)dx = \frac{|P|}{\pi} > 0,$$

which proves the theorem.

2. Sets of multiplicity

The problem of the classification of sets of measure zero into sets U and sets M is far from solved. But it is completely solved for certain families of perfect sets, as we shall show in the next two chapters.

We shall need the following theorem.

THEOREM. *A necessary and sufficient condition for a closed set E to be a set of multiplicity is that there should exist a trigonometric series*

$$\sum_{-\infty}^{\infty} c_n e^{inx}$$

(not vanishing identically) with coefficients $c_n = o\left(\frac{1}{n}\right)$† *and representing a constant*‡ *in each interval contiguous to E.*

Proof. *The condition is necessary.* Let E, closed, be of the type M, and consider a nonvanishing trigonometric series

$$(S) \qquad \sum_{-\infty}^{\infty} \gamma_n e^{nix},$$

converging to zero in every interval contiguous to E.

We show first that we can then construct a series

$$(S') \qquad \sum_{-\infty}^{\infty} \gamma_n e^{nix},$$

but with $\gamma_0 = 0$, having the same property. For (S) has at least one nonvanishing coefficient, say, γ_k. Let $l \neq k$. The series

$$(S') = \gamma_k e^{-ilx}(S) - \gamma_l e^{-ikx}(S)$$

has a vanishing constant term, and converges to zero, like (S), for all x belonging to CE, the complementary set of E. Let E_1 be the set where (S) does not converge to zero. (S') cannot vanish identically, for the only points of E_1 (which is necessarily infinite) where (S') converges to zero are the points (*finite* in number) where

$$\gamma_k e^{-ilx} - \gamma_l e^{-ikx} = 0.$$

Let us then consider the series

$$\sum_{-\infty}^{\infty} \gamma_n e^{nix} \quad (\gamma_0 = 0),$$

converging to zero in CE. The series integrated twice,

$$\sum_{-\infty}^{-1} + \sum_{1}^{\infty} \frac{\gamma_n}{-n^2} e^{inx},$$

represents by Riemann theorems (II and IV on page 43) a linear function in each interval of CE. But this series is the integral of the Fourier series

$$\sum_{-\infty}^{-1} + \sum_{1}^{\infty} \frac{\gamma_n}{ni} e^{inx},$$

which must hence represent a constant in each interval of CE, and it is now enough to remark that

$$c_n = \frac{\gamma_n}{ni} = o\left(\frac{1}{n}\right),$$

since necessarily $\gamma_n \to 0$ (Th. I).

† The series is a Fourier series by the Riesz-Fischer theorem.

‡ Hence, by the elementary theory, converging to this constant.

The condition is sufficient. Suppose that the series

$$\sum_{-\infty}^{\infty} c_n e^{inx}$$

(not vanishing identically) with $c_n = o\left(\frac{1}{n}\right)$ represents a constant in each interval of CE. One can write

$$c_n = \frac{\gamma_n}{ni} \text{ with } \gamma_n \to 0.$$

It follows that the integrated series

$$c_0 x - \sum_{|n| \geq 1} \frac{\gamma_n}{n^2} e^{inx}$$

represents a linear function in each interval of CE. Hence, the series

$$\sum \gamma_n e^{inx}$$

is summable-R to zero in each interval of CE, and thus, by Theorem IIA, converges to zero in each interval contiguous to E, the set E being, therefore, a set of multiplicity.

Remark. If the series

$$\sum_{-\infty}^{\infty} c_n e^{inx}$$

of the theorem represents a function of bounded variation, the series

$$\sum \gamma_n e^{inx}$$

converging to zero in CE is a Fourier-Stieltjes series (in the usual terminology, the Fourier-Stieltjes series of a "measure" whose "support" is E). In this case, we say that E is a set of multiplicity in the restricted sense.

To construct a set of multiplicity in the restricted sense, it is enough to construct a perfect set, support of a measure

$$d\mu \sim \sum_{-\infty}^{\infty} \gamma_n e^{inx},$$

whose Fourier-Stieltjes coefficients,

$$\gamma_n = \frac{1}{2\pi} \int_0^{2\pi} e^{-inx} \, d\mu(x),$$

tend to 0 for $|n| \to \infty$.

Consequence. The results of Chapter IV show that every symmetrical perfect set $E(\xi)$ with constant ratio ξ, such that $1/\xi$ is *not* a number of the class S, is a set of multiplicity. In view of the preceding remark, it is enough to take for μ the Lebesgue function constructed on the set.

3. Construction of sets of uniqueness

We have just seen that in order to show that a closed set E is a set of uniqueness, we must prove that there is no series

$$\sum_{-\infty}^{\infty} c_n e^{nix}$$

(not vanishing identically) with coefficients $c_n = o\left(\frac{1}{n}\right)$ representing a constant in each interval of CE.

We were able to prove only that a symmetrical perfect set $E(\xi)$ is a set M if ξ^{-1} does not belong to the class S, but we cannot, at this stage, prove that if $\xi^{-1} \in S$, then $E(\xi)$ is a set U. This is because we only know that if $\xi^{-1} \in S$, the Fourier-Stieltjes coefficients of the Lebesgue measure constructed on the set do not tend to zero. But we do not know (*a*) whether this is true for *every* measure whose support is $E(\xi)$ or (*b*) whether there does not exist a series

$$\sum_{-\infty}^{\infty} c_n e^{nix}$$

with $c_n = o\left(\frac{1}{n}\right)$ representing a constant in each interval of CE, and which is *not* a function of bounded variation (i.e., the derived series $\sum \gamma_n e^{nix}$ is *not* a Fourier-Stieltjes series).

A negative proof of this kind would be rather difficult to establish. In general, to prove that a set E is a set of the type U, one tries to prove that it belongs to a family of sets of which one knows, by certain properties of theirs, that they are U sets.

In this connection, we shall make use of the following theorem.

THEOREM I. *Let E be a closed set such that there exists an infinite sequence of functions $\{\lambda_k(x)\}_1^{\infty}$ with the following properties:*

1. $\lambda_k(x) = 0$ for all k when $x \in E$.
2. *The Fourier series of each*

$$\lambda_k(x) = \sum_n \gamma_n^{(k)} e^{inx}$$

is absolutely convergent, and we have

$$\sum_n |\gamma_n^{(k)}| < A, \text{ constant independent of } k.$$

3. *We have*

$$\lim_{k=\infty} \gamma_n^{(k)} = 0 \quad \text{for} \quad n \neq 0,$$

$$\lim_{k=\infty} \gamma_0^{(k)} = l \neq 0.$$

4. *The derivative $\lambda_k'(x)$ exists for each x and each k, and is bounded (the bound may depend on k).*

Under these conditions, E is a set of uniqueness.

We shall first prove the following lemma.

LEMMA. *Let E be a closed set, $\lambda(x)$ a function vanishing for $x \in E$ and having an absolutely convergent Fourier series $\sum \gamma_n e^{nix}$, and a bounded derivative $\lambda'(x)$. Let $\sum c_n e^{nix}$ be a trigonometric series converging to zero in every interval of the complementary set CE. Under these conditions we have*

$$\sum \bar{\gamma}_n c_n = 0.$$

(The series is obviously convergent, since $\sum |\gamma_n| < \infty$ and $c_n \to 0$.)

PROOF. Let Δ be an interval contiguous to E. The series

$$c_0 \frac{x^2}{2} - \sum \frac{c_n}{n^2} e^{nix}$$

converges to a linear function in Δ. Hence, the Fourier series

$$f \sim \sum{}^* \frac{c_n e^{nix}}{ni},$$

where the star means that there is no constant term, represents in Δ a function $-c_0 x + a$, the constant $a = a(\Delta)$ depending on Δ. Parseval's formula is applicable [17] in our hypothesis to the functions $f(x)$ and

$$\lambda'(x) \sim \sum \gamma_n n i e^{nix}$$

and gives

$$\frac{1}{2\pi} \int_0^{2\pi} \overline{\lambda'(x)} f(x) dx = - \sum_{|n| \geq 1} \bar{\gamma}_n c_n.$$

The integral is equal to

$$(2\pi)^{-1} \sum_{\Delta} \int_{\Delta} \overline{\lambda'(x)}\, (-c_0 x + a) dx,$$

since λ and λ' are zero for $x \in E$. (Note that if E is closed, but not perfect, its isolated points are denumerable.) Integrating by parts,

$$\int_{\Delta} \overline{\lambda'(x)} (-c_0 x + a) dx = [(-c_0 x + a)\bar{\lambda}]_{\Delta} + c_0 \int_{\Delta} \overline{\lambda(x)} dx,$$

and comparing the three last relations, we have

$$- \sum_{|n| \geq 1} \bar{\gamma}_n c_n = c_0 (2\pi)^{-1} \sum_{\Delta} \int \overline{\lambda(x)} dx$$

$$= c_0 (2\pi)^{-1} \int_0^{2\pi} \overline{\lambda(x)} dx = c_0 \bar{\gamma}_0$$

or, as stated,

$$\sum \bar{\gamma}_n c_n = 0.$$

Remark. The hypothesis that $\lambda'(x)$ is bounded could be relaxed (which would lead also to a relaxation of the hypothesis (4) of the theorem), but this is of no interest for our applications. It should be observed, however, that *some* hypothesis on $\lambda(x)$ is necessary. We know, in fact, since the obtention of recent results on spectral synthesis [6], [8], that the lemma would not be true if we assume only that $\lambda(x) = 0$ for $x \in E$, and that its Fourier series is absolutely convergent

PROOF of Theorem I. Suppose that E is not a set of uniqueness. Hence, suppose the existence of

$$\sum c_n e^{nix}$$

(not identically 0) converging to 0 in each interval of CE. The lemma would then give

$$\sum_n \bar{\gamma}_n^{(k)} c_n = 0 \qquad (3)$$

for all k.

Since $c_n \to 0$ for $n = \infty$ (by general Theorem I on page 43), the hypothesis (2) gives

$$\Big| \sum_{|n| \geq N} \bar{\gamma}_n^{(k)} c_n \Big| < A \cdot \max_{|n| \geq N} |c_n| < A\epsilon,$$

ϵ being arbitrarily small for N large enough. Having fixed N, we have

$$\Big| \sum_{1 \leq |n| < N} \bar{\gamma}_n^{(k)} c_n \Big| < \epsilon$$

for k large enough, by the hypothesis (3) of the theorem.

Hence the first member of (3) differs from $c_0 \bar{l}$ by a quantity arbitrarily small, for k large enough. This proves that $c_0 = 0$.

Multiplying the series

$$\sum c_n e^{nix}$$

by e^{-kix}, we find its constant term to be c_k. Thus the argument gives that $c_k = 0$ for all k, that the series $\sum c_n e^{nix}$ is identically 0, and that E is a set of the type U.

First application: Sets of the type H. A linear set $E \subset (0, 2\pi)$ is said to be "of the type H" if there exists an interval (α, β) contained in $(0, 2\pi)$ and an infinite sequence of integers $\{n_k\}_1^\infty$ such that, for whatever $x \in E$ none of the points of abscissa $n_k x$ (reduced modulo 2π) belongs to (α, β).

For example, the points of Cantor's ternary set constructed on $(0, 2\pi)$:

$$x = 2\pi \left[\frac{\epsilon_1}{3} + \frac{\epsilon_2}{3^2} + \cdots + \frac{\epsilon_k}{3^k} + \cdots \right]$$

where ϵ_j is 0 or 2, form a set of the type H, since the points $3^k x$ (mod 2π) never belong to the middle third of $(0, 2\pi)$. The situation is the same for every symmetrical perfect set $E(\xi)$ with constant ratio ξ, if $1/\xi$ is a rational integer.

THEOREM II. *Every closed set of the type H (and thus also every set of the type H †) is a set U.*

PROOF. Let us fix an $\epsilon > 0$, arbitrarily small and denote by $\lambda(x)$ a function vanishing in $(0, \alpha)$ and in $(\beta, 2\pi)$, equal to 1 in $(\alpha + \epsilon, \beta - \epsilon)$ and having a bounded derivative $\lambda'(x)$, so that its Fourier series is absolutely convergent. Write

$$\lambda(x) = \sum_m \gamma_m e^{mix}$$

and

$$\lambda_k(x) = \lambda(n_k x) = \sum_m \gamma_m e^{mn_k ix}.$$

The sequence of functions $\{\lambda_k(x)\}$ satisfy the conditions (1), (2), (3), (4) of Theorem I. In particular, $\lambda(n_k x)$ is zero for all $x \in E$ and all k, and since

$$\gamma_n^{(k)} = \gamma_m$$

if and only if $n = mn_k$ and $\gamma_n^{(k)} = 0$ if $n_k \nmid n$, we see that the conditions (3) are satisfied, with

$$l = \gamma_0 = \frac{1}{2\pi} \int_0^{2\pi} \lambda(x)dx = (2\pi)^{-1}(\beta - \alpha - 2\epsilon),$$

which is positive if ϵ has been chosen small enough.

Second application. Sets of the type $H^{(n)}$. The sets of the type H have been generalized by Piatecki-Shapiro, who described as follows the sets which he calls "of the type $H^{(n)}$."

DEFINITION. *Consider, in the n-dimensional Euclidean space R^n, an infinite family of vectors $\{V_k\}$ with rational integral coordinates*

$$V_k = \{p_k^{(1)}, p_k^{(2)}, \ldots, p_k^{(n)}\} \quad (k = 1, 2, \ldots).$$

This family will be called normal, *if, given n fixed arbitrary integers $a_1, a_2, \ldots, a_n$ not all zero, we have*

$$| a_1 p_k^{(1)} + a_2 p_k^{(2)} + \cdots + a_n p_k^{(n)} | \to \infty$$

as $k \to \infty$.

Let Δ be a domain in the n-dimensional torus

$$0 \leq x_j < 2\pi \quad (j = 1, 2, \ldots, n).$$

A set E will be said to belong to the type $H^{(n)}$ if there exists a domain Δ and a normal family of vectors V_k such that for all $x \in E$ and all k, the point with coordinates

$$p_k^{(1)}x, p_k^{(2)}x, \ldots, p_k^{(n)}x,$$

all reduced modulo 2π, never belongs to Δ.

† If E is of the type H, so is its closure, and a subset of a U-set is also a U-set.

THEOREM III. *Every set E of the type $H^{(n)}$ is a set of uniqueness.*

PROOF. We can again suppose that E is closed, and we shall take $n = 2$, the two-dimensional case being typical. Suppose that the family of vectors

$$V_k = (p_k, q_k)$$

is normal. We can assume that Δ consists of the points (x_1, x_2) such that

$$\alpha_1 < x_1 < \beta_1,$$
$$\alpha_2 < x_2 < \beta_2,$$

the intervals (α_1, β_1) and (α_2, β_2) being contained in $(0, 2\pi)$.

We shall denote by $\lambda(x)$ and $\mu(x)$ respectively two functions constructed with respect to the intervals (α_1, β_1) and (α_2, β_2) as was, in the case of sets H, the function $\lambda(x)$ with respect to (α, β). Under these conditions, the functions

$$\lambda(p_k x)\mu(q_k x) \quad (k = 1, 2, \ldots)$$

are equal to zero for all k and all $x \in E$. This sequence of functions will play the role of the sequence denoted by $\lambda_k(x)$ in Theorem I. Thus, the condition (1) of that theorem is satisfied.

Write

$$\lambda(x) = \sum \gamma_m e^{imx}, \quad \mu(x) = \sum \delta_m e^{imx}.$$

The Fourier series of $\lambda(p_k x)\mu(q_k x)$ is absolutely convergent, and, writing

$$\lambda(p_k x)\mu(q_k x) = \sum c_n^{(k)} e^{inx},$$

we have

$$\sum c_n^{(k)} e^{inx} = \sum \gamma_m \delta_{m'} e^{i(mp_k + m'q_k)x}$$

and

$$\sum | c_n^{(k)} | < (\sum | \gamma_n |)(\sum | \delta_n |) < A.$$

This proves that condition (2) is also satisfied.

Condition (4) is satisfied if we have chosen $\lambda(x)$ and $\mu(x)$ possessing bounded derivatives.

Finally, for condition (3) we note that

$$c_n^{(k)} = \sum_{n = mp_k + m'q_k} \gamma_m \delta_{m'}. \tag{4}$$

Suppose first $n = 0$. Then

$$c_0^{(k)} = \sum_{mp_k + m'q_k = 0} \gamma_m \delta_{m'}$$
$$= \gamma_0 \delta_0 + \sum_{mp_k + m'q_k = 0}^{*} \gamma_m \delta_{m'} = \gamma_0 \delta_0 + T$$

the star meaning that $| m | + | m' | \neq 0$. We shall prove that T tends to zero for $k \to \infty$. Write $T = T_1 + T_2$, where T_1 is extended to the indices $| m | \leq N$,

$|m'| \leq N$. Since the family of vectors $\{V_k\}$ is normal, if $|m|+|m'| \neq 0$, $mp_k + m'q_k$ cannot be zero if k is large enough, and if m and m' are chosen among the finite number of integers such that $|m| \leq N$, $|m'| \leq N$. On the other hand, in T_2 either $|m| > N_1$ or $|m'| > N$ and thus

$$|T_2| < (\sum_{|n|>N} \gamma_m)(\sum_{-\infty}^{\infty} |\delta_{m'}|) + (\sum_{-\infty}^{\infty} |\gamma_m|)(\sum_{|m'|>N} |\delta_{m'}|)$$

is arbitrarily small for N large enough. Choosing first N, and then k, we see that

$$c_0^{(k)} \longrightarrow \gamma_0\delta_0$$

as $k \rightarrow \infty$, and since $\gamma_0\delta_0 \neq 0$, the second part of condition (3) is satisfied.

If now $n \neq 0$, the second member of (4) does not contain the term where $m = 0$, $m' = 0$. The same argument leads then to

$$c_n^{(k)} \longrightarrow 0 \quad \text{for} \quad k = \infty, n \neq 0.$$

This concludes the proof that all conditions of the general theorem are satisfied and hence that the set E is a set of uniqueness.

In the following two chapters we shall apply the preceding theorems to special sets: symmetrical perfect sets with constant ratio of dissection, and "homogeneous sets."

SYMMETRICAL PERFECT SETS WITH CONSTANT RATIO OF DISSECTION; THEIR CLASSIFICATION INTO M-SETS AND U-SETS

In this chapter and in the following one we shall make use of the fundamental theorem of Minkowski on linear forms. For the proof we refer the reader to the classical literature. (See, e.g., [5].)

MINKOWSKI'S THEOREM. *Consider n linear forms of n variables*

$$\lambda_p(x) = \sum_{q=1}^{n} a_q{}^p x_q \quad (p = 1, 2, \ldots, n)$$

where we suppose first the coefficients $a_q{}^p$ *to be real. We assume that the determinant D of the forms is not zero. If the positive numbers* $\delta_1, \delta_2, \ldots, \delta_n$ *are such that*

$$\delta_1\delta_2 \cdots \delta_n \geq |D|,$$

there exists a point x with rational integral coordinates $(x_1, x_2, \ldots, x_n)$ *not all zero such that*

$$|\lambda_p(x)| \leq \delta_p \quad (p = 1, 2, \ldots, n).$$

The theorem remains valid if the coefficients $a_q{}^p$ are complex numbers provided:

1. the complex forms figure in conjugate pairs
2. the δ_p corresponding to conjugate forms are equal.

THEOREM. *Let* $E(\xi)$ *be a symmetrical perfect set in* $(0, 2\pi)$ *with constant ratio of dissection* ξ. *A necessary and sufficient condition for* $E(\xi)$ *to be a set of uniqueness is that* $1/\xi$ *be a number of the class S* [14].

PROOF. The necessity of the condition follows from what has been said in the preceding chapter. We have only to prove here the sufficiency: If ξ^{-1} belongs to the class S, $E(\xi)$ is a U-set.

We simplify the formulas a little by constructing the set $E(\xi)$ on $[0, 1]$. We write $\theta = 1/\xi$ and suppose, naturally, that $\theta > 2$. We assume that θ is an algebraic integer of the class S and denote by n its degree. We propose to show that $E(\xi)$ is of the type $H^{(n)}$, and hence a set of uniqueness.

The points of $E(\xi)$ are given by

$$x = \epsilon_1 r_1 + \epsilon_2 r_2 + \cdots + \epsilon_j r_j + \cdots$$

where $r_j = \xi^{j-1}(1 - \xi) = \dfrac{1}{\theta^{j-1}}\left(1 - \dfrac{1}{\theta}\right) = \dfrac{\theta - 1}{\theta^j}$ and the ϵ_j are 0 or 1.

Thus,

$$x = (\theta - 1)\left[\frac{\epsilon_1}{\theta} + \frac{\epsilon_2}{\theta^2} + \cdots + \frac{\epsilon_j}{\theta^j} + \cdots\right].$$

By λ we denote a positive algebraic integer of the field of θ, which we shall determine later. We denote by $\alpha_1, \alpha_2, \ldots, \alpha_{n-1}$ the conjugates of θ and by $\mu_1, \mu_2, \ldots, \mu_{n-1}$ the conjugates of λ.

We have, x being a fixed point in $E(\xi)$ and m a rational integer ≥ 0,

$$\lambda\theta^m x = \lambda(\theta - 1)\left(\frac{\epsilon_{m+1}}{\theta} + \cdots\right) + R \tag{1}$$

$$R = \lambda(\theta - 1)(\epsilon_1\theta^{m-1} + \epsilon_2\theta^{m-2} + \cdots + \epsilon_m).$$

Observe that, for any natural integer $p \geq 0$,

$$\lambda(\theta - 1)\theta^p + \sum_{i=1}^{n-1} \mu_i(\alpha_i - 1)\alpha_i^p \equiv 0 \pmod 1.$$

That is to say

$$\lambda(\theta - 1)\theta^p \equiv -\sum_{i=1}^{n-1} \mu_i(\alpha_i - 1)\alpha_i^p.$$

Hence, remembering that the $|\alpha_i|$ are < 1,

$$|R| < 2\sum_{i=1}^{n-1} |\mu_i| \sum_{m=0}^{\infty} |\alpha_i|^m = 2\sum_{i=1}^{n-1} \frac{|\mu_i|}{1 - |\alpha_i|} \pmod 1. \tag{2}$$

Let us now write (1), after breaking the sum in parenthesis into two parts, as

$$\lambda\theta^m x = \lambda(\theta - 1)\left(\frac{\epsilon_{m+1}}{\theta} + \cdots + \frac{\epsilon_{m+N}}{\theta^N}\right) + \lambda(\theta - 1)\left(\frac{\epsilon_{m+N+1}}{\theta^{N+1}} + \cdots\right) + R \tag{3}$$

$$= P + Q + R.$$

We have

$$|Q| < \lambda(\theta - 1)\frac{\theta^{-N-1}}{1 - \theta^{-1}} = \frac{\lambda}{\theta^N}. \tag{4}$$

We now choose λ of the form

$$\lambda = x_1 + x_2\theta + \cdots + x_n\theta^{n-1},$$

where the x_j are rational integers. Then, obviously,

$$\mu_i = x_1 + x_2\alpha_i + \cdots + x_n\alpha_i^{n-1} \quad (i = 1, 2, \ldots, n - 1).$$

By Minkowski's theorem, we determine the rational integers, such that

$$\frac{\lambda}{\theta^n} \leq \frac{\sigma}{n2^{N/n}}; \quad \frac{2|\mu_i|}{1 - |\alpha_i|} \leq \frac{\sigma}{n2^{N/n}} \quad (i = 1, 2, \ldots, n - 1), \tag{5}$$

where σ will be determined in a moment. The determinant of the forms

$$\frac{\lambda}{\theta^N} \quad \text{and} \quad \frac{2\mu_i}{1 - |\alpha_i|} \quad (i = 1, 2, \ldots, n = 1)$$

can be written as

$$\frac{\Delta}{\theta^N},$$

where Δ is a nonvanishing determinant depending only on θ (and independent of N), say, $\Delta = \Delta(\theta)$. Minkowski's theorem can be applied, provided

$$\frac{\sigma^n}{n^n 2^N} > \frac{\Delta}{\theta^N},$$

and, after choosing σ, we can always determine N so that this condition be fulfilled, since $\theta/2 > 1$.

By (2), (3), (4), and (5), we shall then obtain for an arbitrary fixed $x \in E(\xi)$ and any arbitrary natural integer $m \geq 0$

$$|\lambda\theta^m x - P| \leq \frac{\sigma}{2^{N/n}} \pmod 1,$$

that is to say

$$\text{(6)} \qquad \left| \lambda\theta^m x - \lambda(\theta - 1)\left(\frac{\epsilon_{m+1}}{\theta} + \cdots + \frac{\epsilon_{m+N}}{\theta^N}\right) \right| \leq \frac{\sigma}{2^{N/n}} \pmod 1.$$

Denote now by g_m the fractional part of P (depending on m), and denote by O_k, k an arbitrary natural integer, the point having the coordinates $g_{k+1}, g_{k+2}, \ldots, g_{k+n}$.

The *number* of points O_k depends evidently on k, n, and the choice of the ϵ's; but we shall prove that there are at most 2^{N+n-1} distinct points O_k. In fact, observe that g_{k+1} can take 2^N values (according to the choice of the ϵ's). But, once g_{k+1} is fixed, g_{k+2} can only take 2 different values; and, once g_{k+1} and g_{k+2} are fixed, g_{k+3} can take only 2 distinct values. Thus the number of points O_k is at most 2^{N+n-1}.

Let now M_k be the point whose coordinates are

$$(\lambda\theta^{k+1}x), (\lambda\theta^{k+2}x), \ldots, (\lambda\theta^{k+n}x),$$

where (z) denotes, as usual, the fractional part of z. This point considered as belonging to the n-dimensional unit torus is, by (6), interior to a cube of side

$$\frac{2\sigma}{2^{N/n}}$$

and of center O_k. The number of cubes is at most 2^{N+n-1} and their total volume is

$$2^{N+n-1}\frac{(2\sigma)^n}{2^N} = 2^{2n-1}\sigma^n = \frac{1}{2}(4\sigma)^n.$$

If we take $\sigma \leq \frac{1}{4}$, there will remain in the torus $0 \leq x_j < 1$ $(j = 1, 2, \ldots, n)$ a "cell" free of points M_k. This will also be true, for every $k > k_0$ large enough, for the point M_k' of coordinates

$$(c_{k+1}x), \ldots, (c_{k+n}x),$$

if we denote generally by c_m the integer nearest to $\lambda\theta^m$, since we know that $\lambda\theta^m = c_m + \delta_m$ with $\delta_m \to 0$ $(m \to \infty)$.

To show now that $E(\xi)$ is of the type $H^{(n)}$, we have only to prove that the sequence of vectors

$$V_k = (c_{k+1}, c_{k+2}, \ldots, c_{k+n})$$

in the Euclidean space R^n is normal. Let $a_1, a_2, \ldots, a_n$ be natural integers, not all zero. We have

$$a_1c_{k+1} + \cdots + a_nc_{k+n} = \lambda(a_1\theta^{k+1} + \cdots + a_n\theta^{k+n}) + (a_1\delta_{k+1} + \cdots + a_n\delta_{k+n}).$$

If $k \to \infty$, the last parenthesis tends to zero. On the other hand, the first parenthesis equals

$$\lambda\theta^{k+1}(a_1 + a_2\theta + \cdots + a_n\theta^{n-1}),$$

and its absolute value increases infinitely with k, since, θ being of degree n, we have

$$a_1 + a_2\theta + \cdots + a_n\theta^{n-1} \neq 0.$$

This completes the proof.

Remark. We have just proved that if θ belongs to the class S and has degree n the set $E(\xi)$ is of the type $H^{(n)}$. But it does not follow that E cannot be of a simpler type. Thus, for instance, if θ is quadratic, our theorem shows that E is of the type $H^{(2)}$. But in this particular case, one can prove that E is, more simply, of the type H.†

Stability of sets of uniqueness. We have shown in Chapter II that the set of numbers of the class S is closed. If $E(\xi_0)$ is a set M, ξ_0^{-1} belongs to an open interval contiguous to S. Hence, there exists a neighborhood of ξ_0 such that all numbers of this neighborhood give again sets M. Thus, a symmetrical perfect set of the type M presents a certain *stability* for small variations of ξ. On the contrary, if $E(\xi_0)$ is a U-set, there are in the neighborhood of ξ_0 numbers ξ such that $E(\xi)$ is an M-set. The sets of uniqueness are are "stable" for small variations of ξ.

† See *Trans. Amer. Math. Soc.*, Vol. 63 (1948), p. 597.

THE CASE OF GENERAL "HOMOGENEOUS" SETS

1. Homogeneous sets

The notion of symmetrical perfect set with constant ratio of dissection can be generalized as follows.

Considering, to fix the ideas, the interval $[0, 1]$ as "fundamental interval," let us mark in this interval the points of abscissas

$$\eta_0 = 0, \quad \eta_1, \eta_2, \ldots, \eta_d \quad (d \geq 1; \ \eta_d = 1 - \xi),$$

and consider each of these points as the origin of an interval ("white" interval) of length ξ, ξ being a positive number such that

$$\xi < \frac{1}{d+1}$$

$$\eta_{j+1} - \eta_j > \xi \quad (\text{for all } j)$$

so that no two white intervals can have any point in common. The intervals between two successive "white" intervals are "black" intervals and are removed. Such a dissection of $[0, 1]$ will be called of the type $(d, \xi; \ \eta_0, \eta_1, \eta_2, \ldots, \eta_d)$.

We operate on each white interval a dissection homothetic to the preceding one. We get thus $(d+1)^2$ white intervals of length ξ^2, and so on indefinitely. By always removing the black intervals, we get, in the limit, a nowhere dense perfect set of measure zero, whose points are given by

$$x = \epsilon_0 + \epsilon_1 \xi + \epsilon_2 \xi^2 + \cdots, \tag{1}$$

where each ϵ_j can take the values $\eta_0, \eta_1, \ldots, \eta_d$.

The case of the symmetrical perfect set is obtained by taking

$$d = 1, \quad \eta_0 = 0, \ \eta_1 = 1 - \xi.$$

The set E of points (1) will be called "homogeneous" because, as is readily seen, E can be decomposed in $(d+1)^k$ portions, all homothetic to E in the ratio ξ^k $(k = 1, 2, \ldots)$.

2. Necessary conditions for the homogeneous set E to be a U-set

Since each subset of a set of uniqueness is also a set of uniqueness, if we consider the set $E_0 \subset E$ whose points are given by (1) but allowing the ϵ_j to take only the values $\eta_0 = 0$ or $\eta_d = 1 - \xi$, then E_0 is a set U, if E is a set U.

But E_0 is a symmetrical perfect set with constant ratio of dissection ξ. Hence, if the homogeneous set E is a U-set, we have necessarily $\xi = 1/\theta$, where θ is a number of the class S.

Consider further the subset E' of E whose points are given by (1) but with the choice of the ϵ_j restricted as follows:

$$\begin{array}{lll} \epsilon_0 = 0 \text{ or } \eta_1 & \epsilon_d = 0 \text{ or } \eta_1 & \\ \epsilon_1 = 0 \text{ or } \eta_2 & \epsilon_{d+1} = 0 \text{ or } \eta_2 & \text{etc.} \\ \cdots\cdots\cdots & \cdots\cdots\cdots & \\ \epsilon_{d-1} = 0 \text{ or } \eta_d & \epsilon_{2d-1} = 0 \text{ or } \eta_d & \end{array}$$

The points of this set E' are given by

$$x' = \epsilon_1'\eta_1 + \epsilon_2'\eta_2\xi + \cdots + \epsilon_d'\eta_d\xi^{d-1} + \epsilon_{d+1}'\eta_1\xi^d + \cdots = \sum \epsilon_j' r_j,$$

where the ϵ_j' are either 0 or 1.

We can, as in the case of symmetrical perfect sets, define a measure carried by this set and prove that its Fourier-Stieltjes transform is

$$\prod_{k=1}^{\infty} \cos \pi u r_k. \tag{2}$$

If E is a U-set, E' is a U-set and (2) cannot tend to zero if $u \to \infty$. It follows that there exists an infinite sequence of values of u for which each of the infinite products

$$\begin{array}{l} \cos \pi u\eta_1 \cdot \cos \pi u\eta_1\xi^d \cdot \cos \pi u\eta_1\xi^{2d} \cdots \\ \cos \pi u\eta_2\xi \cdot \cos \pi u\eta_2\xi^{d+1} \cdot \cos \pi u\eta_2\xi^{2d+1} \cdots \\ \cdots\cdots\cdots\cdots\cdots\cdots \\ \cos \pi u\eta_d\xi^{d-1} \cdot \cos \pi u\eta_d\xi^{2d-1} \cdot \cos \pi u\eta_d\xi^{3d-1} \end{array}$$

has absolute value larger than a fixed positive number a. Write $\omega = 1/\xi^d$. We have, for an infinite sequence of values of u:

$$\prod_{k=0}^{\infty} |\cos \pi u\eta_h\xi^{h-1} \cdot \xi^{kd}| > a \quad (h = 1, 2, \ldots, d),$$

and from this we deduce, by the same argument as in Chapter IV, the existence of a real number $\Lambda \neq 0$ such that

$$\sum \sin^2 \pi\Lambda\eta_h\xi^{h-1}\omega^n < \infty \quad (h = 1, 2, \ldots, d).$$

We know that from this condition it follows that (1) $\omega \in S$, a condition which we shall suppose to be fulfilled (since we know that we have the necessary condition $\xi^{-1} \in S$, which implies $\xi^{-d} \in S$), (2) the numbers

$$\Lambda\eta_1, \Lambda\eta_2, \ldots, \Lambda\eta_d$$

all belong to the field of ω (hence to the field of $\theta = \xi^{-1}$). Since $\eta_d = 1 - \xi$, it follows that

$$\eta_1, \eta_2, \ldots, \eta_d$$

Summing up our results we get:

THEOREM. *If the homogeneous set E is a set of uniqueness*, then:
1. $1/\xi$ *is an algebraic integer* θ *of the class S.*
2. *The abscissas* $\eta_1, \ldots, \eta_d$ *are algebraic numbers of the field of* θ.

We proceed now to prove that the preceding conditions are *sufficient* in order that E be a U-set.

3. Sufficiency of the conditions

THEOREM. *The homogeneous set E whose points are given by* (1), *where* $1/\xi = \theta$ *is an algebraic integer of the class S and the numbers* $\eta_1, \ldots, \eta_d$ *are algebraic belonging to the field of* θ, *is a set of the type* $H^{(n)}$ (n being the degree of θ), *and thus a set of uniqueness.*

PROOF. Let a be a rational positive integer such that $a\eta_1, a\eta_2, \ldots, a\eta_d$ are integers of the field of θ. Denote by

$$\alpha^{(1)}, \ldots, \alpha^{(n-1)}$$

the conjugates of θ and by

$$\omega_j^{(1)}, \ldots, \omega_j^{(n-1)} \quad (j = 1, 2, \ldots, d)$$

the conjugates of η_j. Denote further by λ an algebraic integer of the field of θ, whose conjugates shall be denoted by

$$\mu^{(1)}, \ldots, \mu^{(n-1)}.$$

Writing (1) in the form

$$x = \epsilon_0 + \frac{\epsilon_1}{\theta} + \frac{\epsilon_2}{\theta^2} + \cdots \quad (\epsilon_j = \eta_0, \eta_1, \ldots, \eta_d)$$

we have, if m is a natural integer ≥ 0,

$$\lambda a\theta^m\eta_j + \sum_{i=1}^{n-1} \mu^{(i)}a \cdot \alpha^{(i)m}\omega_j^{(i)} \equiv 0 \pmod 1.$$

Thus, $x \in E$ being fixed, we have always

$$\lambda a\theta^m x = \lambda a\left(\frac{\epsilon_{m+1}}{\theta} + \cdots + \frac{\epsilon_{m+N}}{\theta^N}\right) + \lambda a\left(\frac{\epsilon_{m+N+1}}{\theta^{N+1}} + \cdots\right) + R \pmod 1,$$

where $N > 1$ is a natural integer to be chosen later on, and where, putting

$$M = \max_{i,j} \{|\omega_j^{(i)}|, \eta_j\}$$

we have

$$R < Ma\sum_{i=1}^{n-1} |\mu^{(i)}| \sum_{m=0}^{\infty} |\alpha^{(i)m}| = Ma\sum_{i=1}^{n-1} \frac{|\mu^{(i)}|}{1 - |\alpha^{(i)}|}.$$

Just as in the case considered in Chapter VI, Minkowski's theorem leads to the determination of the positive algebraic integer λ of the field of θ such that

$$\frac{\lambda a M}{\theta^N(\theta - 1)} < \frac{1}{2n(d+1)^{(N/n)+1}},$$

$$Ma \frac{|\mu_i|}{1 - |\alpha^{(i)}|} \leq \frac{1}{2n(d+1)^{(N/n)+1}},$$

provided that

$$[2n(d+1)^{\frac{N}{n}+1}]^{-n} > |\Delta| \theta^{-N}.$$

Here Δ is a certain nonvanishing determinant depending on the set E and on θ, but not on N. This condition can be written

$$\theta^N > \Delta(d+1)^N[2n(d+1)]^n$$

and will certainly be satisfied for a convenient choice of N, since $\theta > d+1$.

The numbers λ and N being now thus determined, we shall have, for all m and all $x \in E$,

$$\left| \lambda a \theta^m x - \lambda a \left(\frac{\epsilon_{m+1}}{\theta} + \cdots + \frac{\epsilon_{m+N}}{\theta^N} \right) \right| < \frac{1}{2(d+1)^{(N/n)+1}} \quad (\text{mod } 1).$$

The argument is now identical with the one of Chapter VI. It is enough to observe that

$$(d+1)^{N+n-1} \left[\frac{1}{(d+1)^{(N/n)+1}} \right]^n = \frac{1}{d+1}$$

in order to see that there exists in the torus $0 \leq x_j < 1$ ($j = 1, 2, \ldots, n$) a "cell" free of points whose coordinates are the fractional parts of

$$\lambda a \theta^{k+1} x, \lambda a \theta^{k+2} x, \ldots, \lambda a \theta^{k+n} x,$$

the natural integer $k > 0$ and the point $x \in E$ being arbitrary.

Since $\theta \in S$, we have $\lambda a \theta^m = c_m + \delta_m$, c_m being a rational integer and $\delta_m \to 0$. The remainder of the proof is as before, and we observe that the vectors

$$V_k(c_{k+1}, c_{k+2}, \ldots, c_{k+n})$$

form a normal family.

EXERCISE

The notion of symmetric perfect set with constant ratio of dissection (described at the beginning of Chapter IV) can be generalized as follows.

Divide the fundamental interval (say [0, 1]) in three parts of respective lengths ξ_1, $1 - 2\xi_1$, ξ_1 (where $0 < \xi_1 < \frac{1}{2}$). Remove the central part ("black" interval) and divide each of the two "white" intervals left in three parts of lengths proportional to ξ_2, $1 - 2\xi_2$, ξ_2 $(0 < \xi_2 < \frac{1}{2})$. The central parts are removed, and the 4 white intervals left are divided in parts proportional to ξ_3, $1 - 2\xi_3$, ξ_3 $(0 < \xi_3 < \frac{1}{2})$. We proceed like this using an infinite sequence of ratios $\xi_1, \xi_2, \ldots, \xi_n, \ldots$ and we obtain a symmetric perfect set with variable rates of dissection $E(\xi_1, \ldots, \xi_n, \ldots)$.

Suppose now that the sequence $\{\xi_n\}_1^\infty$ is periodic, i.e., that $\xi_{p+j} = \xi_j$ for all j, the period p being a fixed integer. Prove that the set $E(\xi_1, \ldots, \xi_n \ldots)$ can be considered as a "homogeneous set" in the sense of Chapter VII, with a constant rate of dissection

$$X = \xi_1 \cdots \xi_p.$$

Using the results of this chapter, prove that this set is a set of uniqueness if and only if the following hold.

1. X^{-1} belongs to the class S.
2. The numbers $\xi_1, \ldots, \xi_p$ are algebraic and belong to the field of X.

1. The following problem has already been quoted in Chapter I:

Suppose that the real number $\theta > 1$ is such that there exists a real λ with the property that $\| \lambda\theta^n \| \to 0$ as the integer n increases infinitely (without any other hypothesis). Can one conclude that θ belongs to the class S?

Another way to state the same problem is:

Among the numbers $\theta > 1$ such that, for a certain real λ, $\| \lambda\theta^n \| \to 0$ as $n \to \infty$, do there exist numbers θ which are not algebraic?

2. Let us consider the numbers τ of the class T defined in Chapter III. It is known that every number θ of the class S is a limit point of numbers τ (on both sides). Do there exist other limit points of the numbers τ, and, if so, which ones?

3. It has been shown in Chapter IV that the infinite product

$$\Gamma(u) = \prod_{k=0}^{\infty} \cos \pi u \xi^k$$

is, for $0 < \xi < \frac{1}{2}$, the Fourier-Stieltjes transform of a positive measure whose support is a set $E(\xi)$ of the Cantor type and of constant rate of dissection ξ. But this infinite product has a meaning if we suppose only $0 < \xi < 1$, and in the case $\frac{1}{2} < \xi < 1$ it is the Fourier-Stieltjes transform of a positive measure whose support is a whole interval.† We know that $\Gamma(u) = o(1)$ for $u \to \infty$, if and only if ξ^{-1} does not belong to the class S. Let

$$\Gamma_1(u) = \prod_{0}^{\infty} \cos \pi u \xi_1^k,$$

$$\Gamma_2(u) = \prod_{0}^{\infty} \cos \pi u \xi_2^k$$

where ξ_1^{-1} and ξ_2^{-1} both belong to the class S, so that neither $\Gamma_1(u)$ nor $\Gamma_2(u)$ tends to zero for $u = \infty$. What is the behavior of the product

$$\Gamma_1(u) \cdot \Gamma_2(u)$$

as $u \to \infty$? Can this product tend to zero? Example, $\xi_1 = \frac{1}{5}$, $\xi_2 = \frac{1}{7}$.

This may have an application to the problem of sets of multiplicity. In fact, if ξ_1 and ξ_2 are small enough, $\Gamma_1\Gamma_2$ is the Fourier-Stieltjes transform of a measure whose support is a perfect set of measure zero, namely $E(\xi_1) + E(\xi_2)$.† If $\Gamma_1\Gamma_2 \to 0$, this set would be a set of multiplicity.

† See Kahane and Salem, *Colloquium Mathematicum*, Vol. VI (1958), p. 193. By $E(\xi_1) + E(\xi_2)$ we denote the set of all numbers $x_1 + x_2$ such that $x_1 \in E(\xi_1)$ and $x_2 \in E(\xi_2)$.

4. In the case $\frac{1}{2} < \xi < 1$, the measure of which $\Gamma(u)$ is the Fourier-Stieltjes transform can be either absolutely continuous or purely singular.† Determine the values of ξ for which one or the other case arises. (Of course, if $\xi^{-1} \in S$, $\Gamma(u) \neq o(1)$ and the measure is purely singular. The problem is interesting only if ξ^{-1} does not belong to the class S.)

† See Jessen and Wintner, *Trans. Amer. Math. Soc.*, Vol. 38 (1935), p. 48.

APPENDIX

For the convenience of the reader we state here a few definitions and results which are used throughout the book.

We assume that the reader is familiar with the elementary notions of algebraic numbers and algebraic fields. (See, e.g., [5].)

1. An *algebraic integer* is a root of an equation of the form

$$x^k + a_1 x^{k-1} + \cdots + a_k = 0,$$

where the a_j are rational integers, the coefficient of the term of highest degree being 1.

If α is any algebraic number, there exists a natural integer m such that $m\alpha$ be an algebraic integer.

If θ is an algebraic integer of degree n, then the irreducible equation of degree n with rational coefficients, with coefficient of x^n equal to 1, and having θ as one of its roots, has all its coefficients rational integers. The other roots, which are also algebraic integers, are the *conjugates* of θ.

Every symmetric function of θ and its conjugates is a rational integer. This is the case, in particular, for the product of θ and all its conjugates, which proves that it is impossible that θ and all its conjugates have all moduli less than 1.

The algebraic integer θ is a *unit* if $1/\theta$ is an algebraic integer.

2. If (in a given field) $f(x)$ is an irreducible polynomial, and if a root ξ of $f(x)$ is also a root of a polynomial $P(x)$, then $f(x)$ divides $P(x)$ and thus all roots of f are roots of P.

3. If an algebraic integer and all its conjugates have all moduli equal to 1, they are all roots of unity (see [9]).

4. Let R be a ring of real or complex numbers such that 0 is not a limit point of numbers of R. (R is then called a *discontinuous domain of integrity*.) Then the elements of R are rational integers or integers of an imaginary quadratic field (see [9]).

5. There exist only a finite number of algebraic integers of given degree n, which lie with all their conjugates in a bounded domain of the complex plane (see [9]).

6. Let $P(x)$ be a polynomial in a field k. Let K be an extension of k such that, in K, $P(x)$ can be factored into linear factors. If $P(x)$ cannot be so factored in an intermediate field K' (i.e., such that $k \subset K' \subset K$), the field K is said to be a *splitting* field of $P(x)$, and the roots of $P(x)$ generate K.

Let $\alpha_1, \ldots, \alpha_n$ be the roots of $P(x)$ in the splitting field $K = k(\alpha_1, \ldots, \alpha_n)$. Each automorphism of K over k (i.e., each automorphism of K whose restriction to k is the identity) maps a root of $P(x)$ into a root of $P(x)$, i.e., permutes the roots. The group of automorphisms of K over k is called the (Galois) *group* of the equation $P(x) = 0$. This group is a permutation group acting on the roots $\alpha_1, \ldots, \alpha_n$ of $P(x)$.

If $P(x)$ is irreducible in k, the group thus defined is *transitive*.

See, for all this, [1].

7. Uniform distribution modulo 1 of a sequence of numbers has been defined in Chapter I.

A necessary and sufficient condition for the sequence $\{u_n\}_1^\infty$ to be uniformly distributed modulo 1 is that for every function $f(x)$ periodic with period 1 and Riemann integrable,

$$\lim_{n\to\infty} \frac{f(u_1) + \cdots + f(u_n)}{n} = \int_0^1 f(x)dx.$$

H. Weyl has shown that the sequence $\{u_n\}$ is uniformly distributed modulo 1 if and only if for every integer $h \neq 0$,

$$\lim_{n\to\infty} \frac{e^{2\pi i h u_1} + \cdots + e^{2\pi i h u_n}}{n} = 0.$$

In R^p (p-dimensional Euclidean space) the sequence of vectors

$$V_n = (v_n^1, \ldots, v_n^p)$$

is uniformly distributed modulo 1 in the torus T^p, if for every Riemann integrable function

$$f(x) = f(x^1, \ldots, x^p),$$

periodic with period 1 in each x^j, we have

$$\lim_{n\to\infty} \frac{f(V_1) + \cdots + f(V_n)}{n} = \int_{T^p} f(x)dx,$$

the integral being taken in the p-dimensional unit torus T^p.

H. Weyl's criterion becomes

$$\lim \frac{e^{2\pi i(HV_1)} + \cdots + e^{2\pi i(HV_n)}}{n} = 0$$

where (HV_n) is the scalar product

$$h_1 v_n^1 + \cdots + h_n v_n^p$$

and $h_1, \ldots, h_n$ are rational integers not all 0.

If $\omega_1, \omega_2, \ldots, \omega_p$, and 1 are linearly independent, the vector $(n\omega_1, \ldots, n\omega_p)$ is uniformly distributed modulo 1 (see [2]).

8. *Kronecker's theorem.* See [2]. In the form in which we use it in Chapter III it may be stated as follows:

If

$$\theta_1, \theta_2, \ldots, \theta_k, 1$$

are linearly independent, $\alpha_1, \alpha_2, \ldots, \alpha_k$ are arbitrary, and N and ϵ are positive, there exist integers

$$n > N, p_1, p_2, \ldots, p_k$$

such that

$$| n\theta_j - p_j - \alpha_j | < \epsilon \quad (j = 1, 2, \ldots, k).$$

(This may be considered as a weak consequence of the preceding result on uniform distribution modulo 1 of the vector $(n\theta_1, \ldots, n\theta_k)$.)

9. We had occasion to cite *Minkowski's theorem* on linear forms in Chapters I, III, and VI. We restate it here as follows.

Let

$$\lambda_p(x) = \sum_{q=1}^{n} a_q{}^p x_q \quad (p = 1, 2, \ldots, n)$$

be n linear forms of the n variables $x_1, \ldots, x_n$ where the coefficients are real and the determinant D of the forms is not zero. There exists a point x with integral coordinates not all zero, $x_1, \ldots, x_n$ such that

$$| \lambda_p(x) | \leq \delta_p,$$

provided that $\delta_1 \cdots \delta_p \geq | D |$.

The result holds if the coefficients $a_q{}^p$ are complex, provided that complex forms figure in conjugate pairs, and that the two δ_p's corresponding to a conjugate pair are equal.

The theorem is usually proved by using the following result. If K is a convex region of volume V in the Euclidean space R^n with center of symmetry at the origin and if $V > 2^n$, the region K contains points of integral coordinates other than the origin. An extremely elegant proof of this result has been given by C. L. Siegel, *Acta Mathematica*, Vol. 65 (1935).

BIBLIOGRAPHY

[1] Artin, E., *Galois Theory* (Notre Dame Mathematical Lectures, No. 2), 2d ed., rev. Notre Dame, Indiana; University of Notre Dame Press, 1944. See particularly pp. 30 ff., 46, and 76.

[2] Cassels, J. W. S., *An Introduction to Diophantine Approximation.* Cambridge: Cambridge University Press, 1957. See particularly Chapter IV.

[3] Hardy, G. H., *A Course of Pure Mathematics*, 10th ed. Cambridge: Cambridge University Press, 1952. See particularly Chapter VIII, p. 392.

[4] Hardy, G. H., Littlewood, J. E., and Polya, G., *Inequalities*, 2d ed. Cambridge: Cambridge University Press, 1952. See p. 30.

[5] Hecke, E., *Vorlesungen über die Theorie der Algebraischen Zahlen*, 2. Auflage. Leipzig, 1954. See p. 116.

[6] Kahane, J.-P., *Comptes Rendus*,† Vol. 248 (1959), p. 2943.

[7] Koksma, J. F., *Compositio mathematica*, Vol. 2 (1935), p. 250.

[8] Malliavin, P., *Comptes Rendus*,† Vol. 248 (1959), p. 2155.

[9] Polya, G. and Szegö, G., *Aufgaben und Lehrsätze aus der Analysis*, 2 vols. Berlin: Julius Springer, 1925. See Vol. II, pp. 149–150.

[10] Pisot, C., *Annali di Pisa*, Vol. 7 (1938), pp. 205–248.

[11] Pisot, C., and Dufresnoy, J., *Annales Scientifiques de l'École Normale Supérieure*, Vol. 70 (1953), pp. 105–133.

[12] Salem, R., *Duke Mathematical Journal*, Vol. 11 (1944), pp. 103–108.

[13] Salem, R., *Duke Mathematical Journal*, Vol. 12 (1945), pp. 153–172.

[14] Salem, R., and Zygmund, A., *Comptes Rendus*,† Vol. 240 (1955), pp. 2040–2042. See footnotes.

[15] Salem, R., and Zygmund, A., *Comptes Rendus*,† Vol. 240 (1955), pp. 2281–2285. See footnotes.

[16] Weyl, H., *Mathematische Annalen*, Vol. 77 (1916), pp. 313–352.

[17] Zygmund, A., *Trigonometric Series*, 2d ed., 2 vols. Cambridge: Cambridge University Press, 1959. See particularly Vol. I, Chapter VII and Chapter IX.

† *Comptes Rendus Hebdomadaires des Séances de l'Académie des Sciences.*

INDEX

Selected Problems

on

EXCEPTIONAL SETS

by

LENNART CARLESON

PREFACE

In 1959, I prepared a survey of the theory of exceptional sets in a general sense. It was planned to deal with characterizations of "thin" sets by means of capacities, Hausdorff measures, arithmetical conditions etc. and the significance of these concepts to existence problems for harmonic and analytic functions, boundary behaviour, convergence of expansions and to harmonic analysis.

In the meantime, several books have appeared which cover different aspects of the above program. Here should be mentioned:

(1) A. ZYGMUND, Trigonometrical series. 2nd ed. Cambridge 1959;
(2) M. TSUJI, Potential theory in modern function theory. Tokyo 1959;
(3) L. AHLFORS and L. SARIO, Riemann surfaces. Princeton 1960;
(4) G. ALEXITS, Konvergenzprobleme der Orthogonalreihen. Budapest 1960;
(5) J.-P. KAHANE et R. SALEM, Ensembles parfaits et séries trigonométriques. Paris 1963.

In this situation, a survey seemed less desirable. Instead, I have collected here those parts that seemed to contain new or less well-known aspects, methods of proof or results. In this way, the result is no survey but the selected problems only reflect the personal interests of the author. Most results are given for d-dimensional Euclidean space and, quite generally, simplicity has been preferred to generality whenever a conflict has arisen.

The references are collected on p. 99.

Lennart Carleson

TABLE OF CONTENTS

§I. *GENERAL CAPACITIES*

1. *Analytic sets.* Suppose that to every finite set of non-negative integers $(n_1, n_2, \ldots, n_p)$ there is associated a *closed* subset $A_{n_1,n_2,\ldots,n_p}$ of a certain fixed bounded set of d-dimensional Euclidean space. All sets considered are thus uniformly bounded. By means of the "Souslin operation" these sets generate a set a,

$$(1.1) \quad a = \bigcup_{n_1,\ldots,n_p,\ldots} A_{n_1} \cap A_{n_1,n_2} \cap \ldots \cap A_{n_1,\ldots,n_p} \cap \ldots.$$

The sets, which arise under this operation for different choices of $\{A_{n_1,\ldots,n_p,\ldots}\}$, are called *analytic*. We observe that we may assume, whenever this is convenient, that

$$A_{n_1,\ldots,n_p,n_{p+1}} \subset A_{n_1,\ldots,n_p}.$$

The following lemma is easily proved.

LEMMA 1. *The family* F *of analytic sets has the following properties:*

(a) *every closed set belongs to* F;

(b) *if* $a^{(k)}$ *are analytic, so are* $\cap\, a^{(k)}$ *and* $\cup\, a^{(k)}$.

Every Borel set is thus analytic.

Proof. (a) is trivial.

(b) If $A^{(k)}_{n_1,\ldots,n_p}$ generate $a^{(k)}$, the union $\cup\, a^{(k)}$ is obtained under the Souslin operation on the sets $A^{(n_1)}_{n_2,\ldots,n_p}$.

To obtain the intersection we put the pairs (k, p) in a sequence by means of the diagonal process. In accordance with this convention we get for a given sequence $n_1, n_2, \ldots$ the following sequence of sets:

$$A^{(1)}_{n_1},\ A^{(1)}_{n_1, n_2},\ A^{(2)}_{n_3},\ A^{(3)}_{n_4},\ A^{(2)}_{n_3, n_5},\ A^{(1)}_{n_1, n_2, n_6}, \ldots .$$

These sets are called $B_{n_1}, B_{n_1, n_2}, \ldots$. Under the Souslin operation, these sets generate the intersection $\cap a^{(k)}$.

Closed subsets of analytic sets are obtained by means of the following lemma.

LEMMA 2. *We assume that* a *is defined above by means of formula* (1.1). $\{h_i\}$ *is an arbitrary sequence of positive integers. We define*

$$F_p = \bigcup_{\substack{n_i \le h_i \\ i \le p}} A_{n_1} \cap A_{n_1, n_2} \cap \ldots \cap A_{n_1, \ldots, n_p}. \tag{1.2}$$

Then

$$F = \bigcap_p F_p \tag{1.3}$$

is a closed subset of a.

Proof. The definitions show that F_p and hence also F are closed sets. To prove $F \subset a$, choose $x \in F$. To every p, there correspond integers n_{ip}, $n_{ip} \le h_i$, and such that

$$x \in A_{n_{1p}} \cap A_{n_{1p}, n_{2p}} \cap \ldots \cap A_{n_{1p}, \ldots, n_{pp}}.$$

We choose p_ν so that

$$\lim_{\nu \to \infty} n_{ip_\nu} = m_i$$

exist for all i. This is possible since $n_{ip} \le h_i$ and clearly

$$x \in A_{m_1} \cap A_{m_1, m_2} \cap \ldots \cap A_{m_1, \ldots, m_p} \cap \ldots .$$

This implies $x \in a$ and so $F \subset a$.

2. Let $f(E)$ be a non-negative set function, first defined only for compact sets F and then by means of the relation

$$f(E) = \sup_{F \subset E} f(F), \quad F \text{ compact}, \tag{2.1}$$

for all bounded sets E. We also assume

$$f(E_1) \leq f(E_2), \quad \text{if } E_1 \subset E_2. \tag{2.2}$$

Clearly (2.2) need only be assumed for compact sets E. $f(E)$ has the character of an interior measure. We also define an outer measure $f^*(E)$ in the usual way

$$f^*(E) = \inf_{O \supset E} f(O), \quad O \text{ open}. \tag{2.3}$$

This outer measure is assumed to have the following property

$$f^*(E) = \lim_{n \to \infty} f^*(E_n), \quad \text{if } E_n \nearrow E. \tag{2.4}$$

A set function f with the properties (2.1), (2.2) and (2.4) is called a "capacity" and a set E with the property

$$f(E) = f^*(E) \tag{2.5}$$

is called *capacitable.*

THEOREM. *If the compact sets are capacitable, so are all analytic sets.*

Proof. Let E be an analytic set obtained from the family

$A_{n_1,\ldots,n_p}$. Define the set $S^{(h)}$

$$S^{(h)} = \bigcup_{\substack{n_1,n_2,\ldots \\ n_1 \le h}} A_{n_1} \cap A_{n_1,n_2} \cap \ldots A_{n_1,\ldots,n_j} \cap \ldots .$$

Clearly $S^{(h)} \nearrow E$, $h \nearrow \infty$, and by (2.4) we can for any given $\varepsilon > 0$ choose $h = h_1$ so that $S_1 = S^{(h_1)}$ satisfies

$$f^*(S_1) \ge f^*(E) - \frac{\varepsilon}{2}.$$

Let $S_1, \ldots, S_{p-1}$ be defined and consider

$$S_p^{(h)} = \bigcup_{\substack{n_i \le h_i,\ i \le p-1 \\ n_p \le h}} A_{n_1} \cap A_{n_1,n_2} \cap \ldots \cap A_{n_1,\ldots,n_j} \cap \ldots .$$

Then $f^*(S_p^{(h)}) \nearrow f^*(S_{p-1})$ and $h = h_p$ exists so that

$$S_p = S_p^{(h_p)} \text{ satisfies } f^*(S_p) \ge f^*(S_{p-1}) - \varepsilon \cdot 2^{-p}.$$

It follows that

$$(2.6) \qquad f^*(S_p) \ge f^*(E) - \varepsilon, \quad p = 1, 2, \ldots .$$

The sequence h_i being defined we form F_p by means of formula (1.2). The definitions show that $S_p \subset F_p$ and (2.6) and (2.2) imply

$$f^*(F_p) \ge f^*(E) - \varepsilon.$$

F, defined by (1.3), is by Lemma 2 a closed subset of E. On the other hand, let O be an arbitrary open set containing F. Then $O \supset F_p$, if $p \ge p_0$, whence $f(O) \ge f^*(F_{p_0}) \ge f^*(E) - \varepsilon$. Since the closed sets are assumed to be capacitable we find

$$f(E) \ge f(F) = f^*(F) \ge f^*(E) - \varepsilon.$$

The theorem is thus proved.

It should be observed that (2.1), (2.2) and (2.4) do not imply that the closed sets are capacitable as the simple example

$$f(E) = \begin{cases} 1, & \text{if } E \text{ has interior points,} \\ 0 & \text{otherwise} \end{cases}$$

shows.

§ II. *HAUSDORFF MEASURES*

1. Let $h(r)$ be a continuous function defined for $r \geq 0$ with the properties

$$h(0) = 0, \quad h(r) \textit{ increasing}. \tag{1.1}$$

Such a function is called a *measure function*.

Let E be a bounded set. We study all coverings of E with a countable number of spheres S_ν with radii r_ν, that is such that

$$\bigcup_\nu S_\nu \supset E$$

and define

$$M_h(E) = \inf \Sigma\, h(r_\nu)$$

for all such coverings. It is of no importance if S_ν are assumed open or closed. If we also assume $r_\nu \leq \rho$ we get a corresponding lower bound $\Lambda^{(\rho)}$. The limit

$$\Lambda_h(E) = \lim_{\rho \to 0} \Lambda^{(\rho)}$$

clearly exists and is the classical Hausdorff measure of E, possibly and even in general infinite. $M_h(E)$ and $\Lambda_h(E)$ are zero simultaneously. $M_h(E)$ is however more convenient for applications to function theoretic problems.

We shall now introduce an auxiliary function $m_h(E)$. Let G_p denote a net, consisting of intervals with sides 2^{-p}. G_{p+1} is obtained from G_p by subdividing the intervals of G_p into 2^d equal intervals. G is the collection of intervals belonging to any G_p. We cover the set E by intervals $\omega_\nu \in G$, the intervals considered to be closed, and define

$$m_h(E) = \inf \Sigma\, h(\delta_\nu)\,, \tag{1.2}$$

where δ_ν is the side of ω_ν. It is obvious that there exist two positive constants C_1 and C_2, only depending on the dimension of the space, such that

$$C_1\, M_h(E) \leq m_h(E) \leq C_2\, M_h(E). \tag{1.3}$$

2. The importance of the quantity $M_h(E)$ depends on the following theorem.

THEOREM 1. *If $\mu(e)$ is a non-negative set function such that*

$$\mu(S) \leq h(r) \tag{2.1}$$

for every sphere S of radius r, then

$$\mu(E) \leq M_h(E)\,. \tag{2.2}$$

Conversely, there is a constant a, only depending on the dimension, such that for every compact set F, there is a μ with the property (2.1) such that

$$\mu(F) \geq a\, M_h(F)\,. \tag{2.3}$$

Proof. That (2.1) implies (2.2) follows immediately from the definitions. Let $\{S_\nu\}$ satisfy $\cup S_\nu \supset E$. Then

$$\mu(E) \leq \mu(\cup S_\nu) \leq \Sigma\, \mu(S_\nu) \leq \Sigma\, h(r_\nu)\,.$$

For the construction of μ with the property (2.3), we fix an integer n and construct $\mu_n(e)$ such that $\mu(\omega_n) = h(2^{-n})$ for all $\omega_n \in G_n$ for which $\omega_n \cap F \neq \phi$. We let μ have constant density on each $\omega_n \in G_n$. If for some $\omega_{n-1} \in G_{n-1}$,

$$\mu_n(\omega_{n-1}) > h(2^{-n+1}),$$

we reduce the density of μ on the corresponding ω_n's so that the mass on ω_{n-1} becomes $h(2^{-n+1})$. The resulting set function is called μ_{n-1}. We treat μ_{n-1} in a similar way and after n steps we have obtained the set function μ_0. This function has the property

$$\mu_0(\omega_\nu) \leq h(2^{-\nu}), \quad \omega_\nu \in G_\nu, \quad \nu \leq n.$$

The original choice of n determines μ_0, which is now denoted $\mu_0^{(n)}$. Let $n \to \infty$ and choose a weakly convergent subsequence: $\mu_0^{(n_i)} \to \mu$, $i \to \infty$. μ has its support on F and satisfies

$$\mu(\omega_\nu) \leq 2^d\, h(2^{-\nu}), \quad \omega_\nu \in G_\nu. \tag{2.4}$$

On the other hand, let $\bigcup_{j=1}^{N} \omega^{(j)} \supset F$, $\omega^{(j)} \in G$ and $\omega^{(j)} \cap F \neq \phi$. Let n be large enough and consider $\mu_0 = \mu_0^{(n)}$. The reduction from μ_n to μ_0 is such that either $\mu_0(\omega^{(j)}) = h(\delta^{(j)})$ or $\Sigma^* \mu_0(\omega^{(j)}) = h(\delta)$, where the summation Σ^* is taken over all $\omega^{(j)} \subset$ a certain $\omega \in G$ of side δ. Hence the total mass of μ_0 is $\geq \inf \Sigma\, h(\delta_\nu)$, for all *finite* sums. The same inequality then also holds for μ. If we drop the restriction to finite sums we get the smaller lower bound $m_h(F)$. Hence

$$\mu_0(F) \geq m_h(F),$$

which together with (2.4) proves the theorem.

3. $M_h(E)$ is easily seen to be an outer measure:

$$M_h(E) = \inf_{O \supset E} M_h(O)\,, \quad O \text{ open}.$$

This need not hold for the set function $m_h(E)$. It is however clearly true for $d = 1$ and in order to make the argument clear we first consider *linear* sets.

If then $d = 1$ and $h(r)$ is an arbitrary measure function, we thus have

$$\text{(3.1)} \qquad m_h(E) = \inf_{O \supset E} m_h(O).$$

We shall now prove

$$\text{(3.2)} \qquad \lim_{n\to\infty} m_h(E_n) = m_h(E)\,, \quad E_n \nearrow E.$$

Let $\{\varepsilon_n\}$ be a sequence of positive numbers to be specified later and choose a covering $\{\omega_{\nu n}\}$ of E_n such that

$$\text{(3.3)} \qquad \sum_\nu h(\delta_{\nu n}) < m_h(E_n) + \varepsilon_n.$$

For every $x \in E$, let ω be the largest interval $\omega_{\nu n}$ containing x. We also assume, as we may, that $k/2^n \notin E$, for all integers k and n. The different non-intersecting intervals ω —countably many—that we obtain in this way, are denoted ω_μ of lengths δ_μ. Obviously

$$E \subset \cup\,\omega_\mu.$$

We now choose an integer m and consider first those ω_μ's that are taken from $\{\omega_{\nu 1}\}$. They cover a certain subset Q_1 of E_m. The same subset is covered by a certain subsequence of $\{\omega_{\nu m}\}$, denoted $\{\omega_{\nu m}\}^{(1)}$. These intervals are subintervals of the chosen ω_μ's. These intervals ω_μ also cover the smaller set $Q_1 \cap E_1$ and the sum $\Sigma^{(1)} h(\delta_\mu)$ can, by (3.3) for $n = 1$, be diminished by at most ε_1. Hence we find

$$\Sigma^{(1)} h(\delta_\mu) \leq \Sigma^{(1)} h(\delta_{\nu m}) + \varepsilon_1 . \tag{3.4}$$

We then consider ω_μ's taken from $\{\omega_{\nu 2}\}$ but not from $\{\omega_{\nu 1}\}$. As above we find

$$\Sigma^{(2)} h(\delta_\mu) \leq \Sigma^{(2)} h(\delta_{\nu m}) + \varepsilon_2 . \tag{3.5}$$

We repeat the argument until all coverings $\{\omega_{\nu n}\}$, $n \leq m$, have been considered. The m inequalities (3.4), (3.5), . . . , are added which yields

$$\sum_{n=1}^{m} \Sigma^{(n)} h(\delta_\mu) \leq \Sigma\, h(\delta_{\nu m}) + \sum_1^m \varepsilon_\nu \leq m_h(E_m) + \sum_1^m \varepsilon_\nu + \varepsilon_m .$$

We let $m \to \infty$ and find

$$\Sigma\, h(\delta_\mu) \leq \lim_{m\to\infty} m_h(E_m) + \sum_1^\infty \varepsilon_\nu .$$

Since $\sum_1^\infty \varepsilon_\nu$ can be chosen arbitrarily small we find

$$m_h(E) \leq \lim_{m\to\infty} m_h(E_n) .$$

Since the opposite inequality is trivial we have proved (3.2).

If we specialize E_n to be compact and E to be open and bounded (3.2) implies

$$m_h(O) = \sup_{F \subset O} m_h(F) . \tag{3.6}$$

If we choose $f(F) = m_h(F)$ for compact sets F, (3.6) and (3.1) show that $m_h(E) = f^*(E)$ for all sets. (3.2) is then condition (2.4) of section I. (3.1) means that the compact sets are capacitable and hence by the theorem of § I all analytic sets are capacitable. The most interesting consequence of this fact is formulated in the following theorem.

THEOREM 2. *An analytic set of positive Hausdorff measure for some measure function* h *contains a closed subset also of positive* h*-measure.*

So far, we have only proved the result for linear sets. If a $(d-1)$-dimensional hyperplane has h-measure zero, the above argument applies without any change. To obtain the general result, we give the word "covering" a slightly changed meaning: the system $\{\omega\}$ "covers" E if every point of E is an interior point of $\cup\omega$. The corresponding lower bound (1.2) is denoted $m_h'(E)$. It is obvious that the inequality (1.3) holds also for m_h' and that m_h' is an outer measure in the sense of (3.1). An inspection of the proof of (3.2) shows that this relation also is true. These two relations give Theorem 2. We do not carry out the detailed proof.

4. An interesting consequence of Theorem 2 is

THEOREM 3. *Every analytic set* E *of positive Hausdorff measure contains a closed subset such that*

$$0 < \Lambda_h(F) < \infty, \quad F \subset E.$$

Proof. As before, the proof is simpler in the linear case. We therefore assume, for the moment, that $d = 1$.

Suppose that $E \supset F_1 \supset F_2 \supset \ldots \supset F_n$, with F_ν closed, have been constructed such that

$$(4.1) \qquad m_h(F_\nu) = a_\nu, \quad \nu = 1, 2, \ldots, n,$$

and

(4.2) $$\inf \Sigma h(\delta^{(j)}) = b_\nu, \quad \delta^{(j)} \le 2^{-\nu}, \quad \cup\omega^{(j)} \supset F_\nu .$$

By Theorem 2, F_1 with $a_1 > 0$ exists.

Let $m = m(n)$ be a sufficiently large integer to be specified later. Choose a covering of F_n satisfying the conditions of (4.2) and such that

$$\sum_j h(\delta^{(j)}) \le b_n + 2^{-n} .$$

As remarked before, we may assume that the covering is finite $j \le N$. If $\delta^{(1)} < 2^{-n}$, or if $\delta^{(1)} = 2^{-n}$ and $m_h(\delta^{(1)} \cap F_n) < h(2^{-n})$, we do not change F_n inside $\delta^{(1)}$. Otherwise we remove from $\delta^{(1)}$ its subintervals $\omega_m \in G_m$, here considered as open (or half-open for its end-intervals), one at a time and consider the intersection of the remaining closed set with F_n. We must reach a first point when the m_h-measure of this intersection A_1 is smaller than $h(2^{-n})$. Then

(4.3) $$h(2^{-n}) - h(2^{-m}) \le m_h(A_1) < h(2^{-n}) .$$

We treat in a similar way $\delta^{(2)}, \ldots, \delta^{(N)}$ and obtain the sets $A_2, \ldots, A_N$. We now choose

$$F_{n+1} = \bigcup_{\nu=1}^{N} A_\nu .$$

From the definition (4.2) it follows that

(4.4) $$b_{n+1} \le b_n + 2^{-n} .$$

On the other hand, when we consider coverings of F_{n+1} by arbitrary intervals $\omega^{(j)}$, F_{n+1} does not differ from F_n as far as intervals $\omega^{(j)}$ of length $\ge 2^{-n}$ are concerned. For subintervals of $\delta^{(j)}$ we have the inequalities (4.3). Hence

(4.5) $$a_{n+1} \geq a_n - N \cdot h(2^{-m}).$$

Choosing m large enough we get $\lim_{n\to\infty} a_n > 0$ while $\overline{\lim}\, b_n < \infty$. We now define

$$F = \bigcap_{n=1}^{\infty} F_n.$$

Then $\Lambda_h(F) \leq \lim b_n < \infty$ while $m_h(F) > 0$ since we need only consider finite coverings.

The proof goes through without change if E does not intersect the hyperplanes used in the construction of the nets G_n. If an intersection has h-measure zero we may remove it without changing $m_h(E)$. Therefore the only alternative is that $m_h(E \cap H) > 0$ for some hyperplane H. We have then reduced the dimension of the problem and can repeat the considerations. The result is therefore generally valid.

§ III. *POTENTIAL THEORY*

1. Let $H(t)$ be a non-negative, continuous, increasing, convex function of the real variable t, $-\infty < t < \infty$. Let $\phi(r)$, $r = |x|$, be a fundamental solution of Laplace's equation

$$\phi(r) = \begin{cases} \log \frac{1}{r}, & d = 2 \\ r^{2-d}, & d \neq 2. \end{cases}$$

We shall, in order to keep a good balance between generality and simplicity, only study kernels $K(r)$ of the form

$$K(r) = H(\phi(r)).$$

We also assume

$$\int_0 K(r)\, r^{d-1}\, dr < \infty.$$

With respect to $K(r)$ we form the potential of the real completely additive set function σ,

$$u_\sigma(x) = \int K(|x-y|)\, d\sigma(y)$$

and the energy integral

$$I(\sigma) = \iint K(|x-y|) d\sigma(x)\, d\sigma(y).$$

For σ with a variable sign we study $I(\sigma)$ only when $I(|\sigma|) < \infty$.

If we restrict ourselves to non-negative measures μ with bounded support, u and I have the following properties of semi-continuity.

LEMMA 1.

(a) $\underline{\lim}_{x \to x_0} u_\mu(x) \geq u_\mu(x_0)$.

If $\mu_n \to \mu$ *weakly then*

(b) $\underline{\lim}_{n \to \infty} u_{\mu_n}(x) \geq u_\mu(x)$;

(c) $\underline{\lim}\, I(\mu_n) \geq I(\mu)$.

Proof. The relations follow immediately from the definitions.

THEOREM 1 (The maximum principle). *If* $u_\mu(x) \leq 1$ *on the support* S_μ *of* μ, *then* $u_\mu(x) \leq 1$ *everywhere.*

Proof. By Egoroff's theorem there is a closed subset F of S_μ such that, for a given $\varepsilon > 0$,

$$\mu(F) > \mu(S_\mu) - \varepsilon \tag{1.1}$$

and such that $u_\mu(x)$ converges uniformly on F. If we define $\mu_1(e) = \mu(e \cap F)$, then clearly $u_{\mu_1}(x)$ converges uniformly on F, i.e., for any $x_0 \in F$

$$\int_{|y-x_0|<\eta} K(|y-x_0|)\, d\mu_1(y) < \delta\,, \quad \eta = \eta(\delta).$$

Let $\{x_n\}$ be a sequence such that $x_n \to x_0 \in F$. Then

$$\overline{\lim_{n\to\infty}}\, u_{\mu_1}(x_n) \leq \int_{|y-x_0|\geq\eta} K(|y-x_0|)\, d\mu_1(y) + \tag{1.2}$$

$$+ \overline{\lim_{n\to\infty}} \int_{|y-x_n|<\eta} K(|y-x_n|)\, d\mu_1(y).$$

Now there is a number N, only depending on d, with the following property. For any x there are N overlapping closed cones Q_ν with vertices at x such that if ξ_ν is the point of $Q_\nu \cap F$ which is closest to x, then any other point $y \in F$ is closer to some ξ_ν than to x. If ξ_ν are chosen for $x = x_n$, this means that

$$K(|y - x_n|) \leq \sum_{\nu=1}^{N} K(|y - \xi_\nu|, \quad y \in F, \quad \xi_\nu \in F.$$

The last term of (1.2) is thus $\leq N\delta$, and we have proved

$$\overline{\lim_{x \to x_0}} \; u_{\mu_1}(x) \leq u_{\mu_1}(x_0).$$

By Lemma 1(a), $u_{\mu_1}(x)$ is continuous in the entire space and since it is subharmonic outside F, we have $u_{\mu_1}(x) \leq 1$.

Finally, let z be an arbitrary point, $z \notin S_\mu$. Then if ρ is the distance from z to S_μ, we have

$$u_\mu(z) \leq u_{\mu_1}(z) + \varepsilon K(\rho) \leq 1 + \varepsilon K(\rho).$$

We now let $\varepsilon \to 0$ and obtain the desired result.

The proof of Theorem 1 contains the following result.

THEOREM 2 (The continuity principle). *If* $u_\mu(x)$ *is continuous on* S_μ, *then* $u_\mu(x)$ *is continuous everywhere.*

Proof. We need only observe that by Dini's theorem, the integrals are uniformly convergent on S_μ.

2. *Capacity.* Let E be a bounded Borel set and Γ_E the class of distributions of mass on E, i.e., non-negative set functions μ with $S_\mu \subset E$, with the property

$$u_\mu(x) \leq 1, \quad x \in E. \tag{2.1}$$

Equivalently we may assume (2.1) for all x. The capacity $C_K(E)$ of E is defined by the relation

$$\sup_{\mu \in \Gamma_E} \mu(E) = C_K(E). \tag{2.2}$$

A property that holds except on a set of capacity zero, is said to hold p.p. Observe that if u_ν is a bounded potential, then ν vanishes for sets of capacity zero.

The notion of equilibrium is closely related to the concept of capacity. We shall first show the existence of equilibrium distributions in the following form.

THEOREM 3. *Let F be a compact set and assume that for every point $x \in F$ there is a bounded half-cone $V_x \subset F$. If furthermore*

$$\frac{K(r)}{K(2r)} = O(1), \quad r \to 0, \tag{2.3}$$

$\mu \in \Gamma_F$ exists so that $u_\mu(x) \equiv 1$ on F and $\mu(F) = C_K(F)$.

Proof. We study the variational problem

$$\gamma = \inf I(\mu), \quad S_\mu \subset F, \quad \mu(F) = 1.$$

By Lemma 1(c), the lower bound γ is assumed for a certain distribution of unit mass μ. We shall prove that $u_\mu \equiv \gamma$ on F.

1) $u_\mu(x) \geq \gamma$ *on* F *except on a set of capacity zero.*

Let us namely assume that $u_\mu(x) < \gamma - \varepsilon$ on $T \subset F$, $C(T) > 0$. Let τ be a distribution of unit mass on T so that $u_\tau(x) \leq K$ and form

$$\mu_\delta = (1-\delta)\mu + \delta\tau$$

so that $\mu_\delta \in \Gamma_F$ and $\mu_\delta(F) = 1$. We find

$$\begin{aligned} I(\mu_\delta) &\leq I(\mu) - 2\delta\, I(\mu) + 2\delta \int u_\mu \, d\tau + O(\delta^2) \leq \\ &\leq \gamma - 2\delta\gamma + 2\delta(\gamma - \varepsilon) + O(\delta^2) \leq \\ &\leq \gamma - 2\delta\varepsilon + O(\delta^2) < \gamma \end{aligned}$$

if $\delta > 0$ is small enough.

2) *$u_\mu(x) \leq \gamma$ on S_μ and so everywhere.*

Since a set e with $\mu(e) > 0$ has positive capacity, 1) implies that $u_\mu(x) \geq \gamma$ on S_μ except on a set where μ vanishes. If $u_\mu(x_0) > \gamma$, $x_0 \in S_\mu$, then $u_\mu(x) > \gamma$ in a neighbourhood of x_0 which must have positive μ-measure. This contradicts, however, $I(\mu) = \gamma$.

3) *If (2.3) holds and V_x exists, then $u_\mu(x) = \gamma$.*

Let $x = 0$ be such a point in F and let Q be the cone V_x. Let a and b, $a < b$, be real numbers, sufficiently small, and define

$$q(x) = \begin{cases} \dfrac{K(|x|)}{\int_{|y|<|x|} K(|y|)dy}, & x \in Q, \quad a < |x| < b \\ 0 & \text{otherwise}. \end{cases}$$

There exist arbitrarily small numbers a and b so that

$$\int_{-\infty}^{\infty} q(x)dx = 1.$$

By 1) and 2) we find

$$\gamma = \int u_\mu(x)\, q(x)dx = \int_F d\mu(y) \int K(|x-y|)\, q(x)dx. \tag{2.4}$$

Let $\rho > 0$ be a fixed number. When $a, b \to 0$ we have uniformly

$$\int K(|x-y|)\, q(x)dx \to K(|y|), \quad |y| > \rho.$$

Furthermore we have, $|y| \leq \rho$,

$$\int_{a<|x|<b} \frac{K(|x|)\,K(|x-y|)}{\int_{|t|<|x|} K(|t|)dt}\,dx \leq \text{Const.}\; K(|\frac{y}{2}|) \leq \text{Const.}\; K(|y|)$$

by (2.3).

This implies

$$\int_{|y|<\rho} d\mu(y) \int K(|x-y|)q(x)dx \leq \text{Const.} \int_{|y|<\rho} K(|y|)d\mu(y) < \delta(\rho),$$

where $\delta(\rho) \to 0$, $\rho \to 0$, since by 2) $\int K(|y|)d\mu(y) < \infty$. If first $a, b \to 0$, and then $\rho \to 0$ we find $u_\mu(0) = \gamma$.

Let us now choose $\mu_0 = \gamma^{-1}\mu$. Then $u_{\mu_0} \equiv 1$ on F and $\mu_0 \in \Gamma_F$. Finally suppose $\nu \in \Gamma_F$. Then

$$\mu_0(F) \geq \int u_\nu(x)\,d\mu_0(x) = \int u_{\mu_0}(x)\,d\nu(x) = \nu(F).$$

This inequality completes the proof of Theorem 3.

For an arbitrary kernel $K(|x|)$ and an arbitrary compact set we have the following theorem.

THEOREM 4. *For any kernel* $K(|x|)$ *and any compact set* F, *there exists* $\mu \in \Gamma_F$ *so that*

$$u_\mu(x) = 1 \quad \textit{p.p. on} \quad F$$

and

$$\mu(F) = C_K(F).$$

This follows immediately from 1) and 2) in the proof of Theorem 3.

THEOREM 5. *The extremal problems*

$$A^{-1} = \inf_{\nu} I(\nu), \qquad S_\nu \subset F, \ \nu(F) = 1,$$
$$B = \inf_{\nu} \{\text{total mass of } \nu\}, \qquad u_\nu \geq 1 \ \ p.p. \text{ on } F$$
$$C = \sup \nu(F), \qquad u_\nu \leq 1 \ \ p.p. \text{ on } F$$

are equivalent so that

$$A = B = C = C_K(F).$$

Proof. Let μ be the distribution of Theorem 4.

(A) By 1) and 2) in the proof of Theorem 3, a solution ν of the A-problem has the property $u_\nu = A^{-1}$ p.p. on F and $u_\nu \leq A^{-1}$ everywhere. Hence

$$\frac{C_K(F)}{A} = \int u_\nu \, d\mu = \int u_\mu \, d\nu = 1$$

since a distribution with a bounded potential vanishes for all sets of capacity zero.

(B) If $u_\nu \geq 1$ p.p. on F, then

$$C_K(F) \leq \int u_\nu \, d\mu = \int u_\mu \, d\nu \leq \int d\nu .$$

Hence

$$B \geq C_K(F).$$

The opposite inequality is obvious.

(C) If $u_\nu \leq 1$ p.p. on F, then

$$C_K(F) \geq \int_F u_\nu \, d\mu = \int_F u_\mu \, d\nu \geq \nu(F).$$

Again, the opposite inequality is obvious.

THEOREM 6. *The solution of the extremal problems (A) and (C) of Theorem 5 is unique and identical for the two problems.*

We observe that problem (B) need not have a unique solution as this example for $d = 2$ shows:

$$F: x^2 + y^2 = 1, \qquad K(r) = \overset{+}{\log} \frac{2}{r},$$

μ_1: uniform on F; μ_2: mass concentration at $(0, 0)$.

The proof of Theorem 6 is by means of Fourier transformations. It should be observed that it is here an advantage to deal with general kernels, since we may assume $K(r) \equiv 0$, $r > r_0$. The case $K = \log \frac{1}{r}$. $d = 2$, requires a special simple treatment which we omit here. We first prove two lemmas.

LEMMA 2. *Suppose* $K(r) \equiv 0$, $r > r_0$. *Then, for* $d \geq 2$,

$$F(\xi) = \int_{-\infty}^{\infty} K(r) e^{i(x,\xi)} dx > 0, \quad all \ \ \xi, \ r = |x|.$$

For $d = 1$, $F(\xi) = 0$ *can hold at isolated points.*

Proof. Let Σ denote the unit sphere and $d\sigma$ its area-element, normalized so that $\sigma(\Sigma) = 1$. $F(\xi)$ is clearly a function of $|\xi|$ only and to simplify the notations we assume $|\xi| = 1$. We have, c_ν denoting positive constants only depending on d,

$$F(\xi) = c_1 \int_0^{\infty} K(r) r^{n-1} dr \int_{\Sigma} e^{irx_1} d\sigma_x = c_1 \int_0^{\infty} K(r) J(r) r^{n-1} dr, \ |\xi| = 1.$$

The function $J(t)$ can be written

$$J(t) = \begin{cases} c_2 \operatorname{Re} \{\int_0^{\pi/2} e^{it\cos\phi} \sin^{d-2}\phi \, d\phi\}, & d \geq 2 \\ \cos t, & d = 1. \end{cases}$$

It is easy to prove (and of course well known) that $J(t)$ has the following properties:

$$\begin{cases} J(0) = 1, \quad J'(0) = 0; \\ J(t) < 1; \quad t \neq 0, \quad d \geq 2; \quad t \neq 2\pi n, \quad d = 1; \\ J(t) = -J''(t) - \frac{d-1}{t} J'(t). \end{cases}$$

If the last expression is inserted in the formula for $F(\xi)$, $|\xi| = 1$, we get, using the first property of J in the partial integrations,

$$F(\xi) = -c_1 \int_0^\infty K(r) [(d-1)r^{d-2}J' + r^{d-1}J''] \, dr =$$

$$= c_1 \int_0^\infty K' r^{d-1} J' dr = c_1 \int_0^\infty (1-J) \, d(K' r^{d-1}).$$

The convexity assumption on K is equivalent to $d(K'r^{d-1}) \geq 0$. By the second property of J and the fact that $d(K'r^{d-1}) \equiv 0$ is excluded we get the assertion for $d \geq 2$. The case $d = 1$ is easily proved.

LEMMA 3. *For all kernels K considered and all $\sigma \not\equiv 0$ with compact support such that $I(|\sigma|) < \infty$, $I(\sigma) > 0$ holds.*

Proof. Since σ has compact support we may assume that $K(r) \equiv 0$, $r > r_0$. We write

$$I(\sigma) = \int u(x) \, d\sigma(x), \quad u(x) = \int K(|x-y|) \, d\sigma(y).$$

Denoting the (suitably normalized) Fourier transforms of u, K and $d\sigma$ $\hat{u}, \hat{K}$, and $\hat{\sigma}$ respectively, we find $\hat{u}(\xi) = \hat{K}(\xi)\,\hat{\sigma}(\xi)$ and—formally—by Parseval's relation

$$I(\sigma) = \int \hat{K}(\xi) \, |\hat{\sigma}(\xi)|^2 \, d\xi.$$

To justify this formula, let σ_1 be a restriction of σ so that

$|I(\sigma) - I(\sigma_1)| < \varepsilon$ and u_{σ_1} is continuous. To see that σ_1 exists, set $\sigma - \sigma_1 = \tau$. Then

$$|I(\sigma) - I(\sigma_1)| \leq |2 \int u_{\sigma_1} d\tau| + |I(\tau)| < 3 \int u_{|\sigma|}\, d|\tau|$$

and the last expression tends to zero with $\int d|\tau|$. Let $\phi_n(x)$ be $\gamma_n \exp\{-n|x|^2\}$, normalized so that $\int \phi_n(x)dx = 1$. Then $\hat{\phi}_n > 0$ and clearly by Parseval's formula

$$\int \hat{\phi}_n(\xi)\, \hat{K}(\xi)\, |\hat{\sigma}_1(\xi)|^2 d\xi = \int \phi_n * u_{\sigma_1}(x)\, d\sigma_1(x).$$

Since by Lemma 1 $\hat{\phi}_n \hat{K} \geq 0$ we can first let $n \to \infty$ and then $\varepsilon \to 0$ and obtain the desired formula for $I(\sigma)$.

By Lemma 1 $I(\sigma) \geq 0$. If $I(\sigma) = 0$ we must have $\hat{\sigma}(\xi) \equiv 0$, that is $\sigma \equiv 0$.

Proof of Theorem 6.

(A) Every solution of $\inf I(\mu)$, $\mu(F) = 1$, has by Theorem 2 the properties

$$u_\mu \leq C_K(F)^{-1} \quad \text{everywhere}$$

$$u_\mu = C_K(F)^{-1} \quad \text{p.p. on } F.$$

Let μ_1 and μ_2 be two extremals. Then

$$I(\mu_1 - \mu_2) = I(\mu_1) - 2 \int u_{\mu_1} d\mu_2 + I(\mu_2) = 0,$$

since μ_2 vanishes on the set where $u_{\mu_1} < C_K(F)^{-1}$.

(C) Assume that $u_\nu \leq 1$ and that $\nu(F) = C_K(F)$. Let μ be the unique solution of (A). Then

$$I(\mu - \nu) = \int u_\nu d\nu - 2 \int u_\mu d\nu + I(\mu) \leq$$
$$\leq C_K(F) - 2\, C_K(F) + C_K(F) = 0.$$

3. The definition shows that $C_K(E)$ satisfies the relations (2.1) and (2.2) of section I. The corresponding outer measure (I.2.3) is denoted $C_K^*(E)$. We shall show that the conditions of Theorem I are fulfilled so that the following theorem holds.

THEOREM 7. *Every analytic set is capacitable for the set function* $C_K(E)$.

Proof. 1) *The compact sets are capacitable for* $C_K(E)$.

Let F be a compact set and let O_n be the set of points x with distance $< n^{-1}$ to F. Let $\mu_n \in \Gamma_{O_n}$ be such that $\mu_n(O_n) > C_K(O_n) - n^{-1}$. Since $\mu_n(O_n)$ clearly are bounded, there is a weakly convergent sequence $\mu_{n_\nu} \to \mu$. By Lemma 1 (b), $\mu \in \Gamma_F$. Since

$$\mu(F) = \lim_{\nu \to \infty} \mu_{n_\nu}(O_{n_\nu}) = \lim_{n \to \infty} C_K(O_n),$$

we have

$$C_K(F) \geq C_K^*(F).$$

2) *If* $E_n \nearrow E$, *then* $C_K^*(E_n) \nearrow C_K^*(E)$.

The proof of this statement requires a number of lemmas. All sets used are assumed to be contained in a fixed bounded set M.

LEMMA 4. *For any sets* E_n

$$C_K\left(\bigcup_{n=1}^{\infty} E_n\right) \leq \sum_1^{\infty} C_K(E_n).$$

Proof. Choose $\mu \in \Gamma_{\cup E_n}$. We can write $\mu = \sum_1^{\infty} \mu_n$ where μ_n is the restriction of μ to E_n or, if $\{E_n\}$ have common points, to a subset of E_n. Clearly $\mu_n \in \Gamma_{E_n}$, which yields the lemma.

LEMMA 5. *For any sets* E_n

$$C_K^*(\bigcup_{n=1}^{\infty} E_n) \leq \sum_{n=1}^{\infty} C_K^*(E_n).$$

Proof. Choose $O_n \supset E_n$ so that $C_K(O_n) \leq C_K^*(E_n) + \varepsilon \cdot 2^{-n}$. By Lemma 2

$$C_K(\bigcup_1^{\infty} O_n) \leq \sum_{n=1}^{\infty} C_K^*(E_n) + \varepsilon.$$

The lemma is proved.

LEMMA 6. *If* $u_\mu(x) \leq 1$, *there exists for any* $\varepsilon > 0$ *an open set* O *such that* $C_K(O) < \varepsilon$ *and such that* $u_\mu(x)$ *is continuous on the complement* O' *of* O.

Proof. By the proof of Theorem 1, there exists for any $\delta > 0$ a restriction μ_1 of μ so that $u_{\mu_1}(x)$ is continuous and

$$u_\mu(x) = u_{\mu_1}(x) + u_\nu(x)$$

and

$$\nu(M) < \delta.$$

The set S_n where $u_\nu(x) > n^{-1}$ is by Lemma 1 (a) open and

$$C^*(S_n) = C(S_n) \leq n\,\delta,$$

for if $\lambda \in \Gamma_{S_n}$ we have

$$\delta > \int u_\lambda \, d\nu = \int u_\nu \, d\lambda > n^{-1} \lambda(S_n).$$

Choose n_i and δ_i so that $\sum_i n_i \delta_i < \varepsilon$. The set

$$O = \bigcup_i S_{n_i}$$

is open and by Lemma 2 $C(O) < \varepsilon$. Since for every $\delta > 0$ $u_{\mu_1}(x)$ is continuous, it follows that the oscillation of $u_\mu(x)$ at a point outside S_n is $< n^{-1}$. Hence $u_\mu(x)$ is continuous outside O.

LEMMA 7. *If $\mu_n \to \mu$ weakly, then*

$$\lim_{n\to\infty} u_{\mu_n}(x) = u_\mu(x)$$

except in a set of exterior capacity zero.

Proof. Given $\varepsilon > 0$, there is by Lemmas 4 and 6 an open set O so that u_{μ_n}, $n = 1, 2, \ldots,$ and u_μ are continuous outside O and $C_K(O) < \varepsilon$. Let r and ρ be rational numbers, $r < \rho$, and consider the set F_n:

$$F_n = \{x \mid u_{\mu_n}(x) \ge r, \quad u_\mu(x) \le \rho, \quad x \in O'\}.$$

F_n is closed and so is then

$$\Phi_n = \bigcap_{\nu=n}^{\infty} F_\nu .$$

We shall first show that

$$C_K(\Phi_n) = 0. \tag{3.1}$$

If (3.1) does not hold, there exists $\nu \in \Gamma_{\Phi_n}$ so that $\nu \not\equiv 0$ and $u_\nu(x)$ is continuous for all x. We find

$$0 = \lim_{n\to\infty} \int u_\nu(x)\, d(\mu - \mu_n) = \lim_{n\to\infty} \int (u_\mu(x) - u_{\mu_n}(x))\, d\nu(x) \ge (\rho - r)\,\nu(\Phi_n).$$

This contradiction proves (3.1).

It now follows from (3.1) and Lemma 5 that

$$C_K^*(\bigcup_{r,\rho} \bigcup_n \Phi_n) \le \sum_{r,\rho,n} C_K^*(\Phi_n) = 0 .$$

The set $\bigcup_{r,\rho,n} \Phi_n$ contains however the exceptional set of Lemma 7 and this lemma is thus proved.

LEMMA 8. *To an open set* O *there corresponds a mass distribution* μ *such that*

(a) $u_\mu = 1$ *on* O *except on a set of exterior capacity zero.*

(b) $u_\mu \leq 1$

(c) $\mu(M) = C_K(O)$.

Proof. Let $F_n \nearrow O$ and suppose that $u_{\mu_n} = 1$ p.p. on F_n and $\mu_n(F_n) \nearrow C_K(O)$, $\mu_n \to \mu$. The subset of F_n where $u_{\mu_n} \leq 1 - k^{-1}$ is a closed set of capacity zero and hence by 1) of exterior capacity zero. Since this holds for every k, it follows from Lemma 5 that $u_{\mu_n} = 1$ except on a set of exterior capacity zero. We now use Lemma 7 and again Lemma 5 and deduce that μ has all properties (a), (b) and (c) of Lemma 8.

We can now complete the proof of Theorem 7. Let $E_n \nearrow E$ and choose $O_n \supset E_n$ and with O_n also μ_n of Lemma 8.

$$\mu_n(M) = C_K(O_n) \leq C_K^*(E_n) + n^{-1}.$$

We assume that $\mu_n \to \mu$ so that, by Lemma 7, $u_\mu = 1$ on E except on a set S of exterior capacity zero and $\mu_n(M) \leq \lim C_K^*(E_n)$. Let O_ε be the open set where $u_\mu > 1 - \varepsilon$, $O_\varepsilon \supset E - S$. Hence

$$C_K^*(E) \leq C_K^*(E - S) + C_K^*(S) \leq C_K(O_\varepsilon) \leq \frac{\mu(M)}{1-\varepsilon} \leq$$

$$\leq \lim_{n\to\infty} C_K^*(E_n) \cdot (1 - \varepsilon)^{-1}.$$

Thus

$$C_K^*(E) \leq \lim_{n\to\infty} C_K^*(E_n).$$

§ IV. *CERTAIN PROPERTIES OF HAUSDORFF MEASURES AND CAPACITIES*

1. There is a close connexion between Hausdorff measures and capacities which is exhibited by the following theorem.

THEOREM 1. *For any bounded set* E, $C_K(E) > 0$ *implies* $\Lambda_{\overline{K}-1}(E) = \infty$, *where*

$$\overline{K}(r) = r^{-d} \int_0^r K(t)\, t^{d-1}\, dt \, .$$

If conversely E *is analytic and* $\Lambda_h(E) > 0$ *for a measure function* h *such that*

$$\int_0 K(r)\, dh(r) < \infty \, , \tag{1.1}$$

then $C_K(E) > 0$.

Proof. 1) Assume $C_K(E) > 0$, so that $\mu \not\equiv 0$ concentrated on $F \subset E$ exists with $u_\mu(x) \leq 1$. The restriction μ_1 of μ to a suitable $F_1 \subset F$, $\mu_1 \not\equiv 0$, corresponds to a uniformly continuous potential u_{μ_1}. Then, by Dini's theorem, there exists another kernel $K_1(r)$ such that

$$\lim_{r \to 0} \frac{K(r)}{K_1(r)} = 0 \tag{1.2}$$

and such that

$$\int K_1(|x - y|)\, d\mu_1(y) \leq M \tag{1.3}$$

for all x. Now let $\{S_\nu\}$ be a finite covering of F_1 by open spheres $|x - x_\nu| < r_\nu$ and assume $r_\nu \leq \rho$. Then forming mean-values over spheres with twice the radius we find, by (1.3), for a certain constant M_1,

$$\mu_1(F_1) \leq \Sigma\, \mu_1(S_\nu) \leq \Sigma\, \frac{M_1}{\overline{K}_1(r_\nu)} \leq M_1 \sup_{r \leq \rho} \frac{\overline{K}(r)}{\overline{K}_1(r)} \sum_\nu \frac{1}{\overline{K}(r_\nu)} .$$

We now use (1.2). The above inequality then implies for a certain $\varepsilon(\rho) \to 0$, $\rho \to 0$,

$$\Lambda^{(\rho)}_{\overline{K}-1} \geq \frac{\mu_1(F)}{M_1 \varepsilon(\rho)} .$$

Letting $\rho \to 0$ we find $\Lambda_{\overline{K}-1}(E) = \infty$ as asserted.

2) Since E is analytic there is a closed set F with $M_h(F) > 0$. Hence, by Theorem II.1, there is a mass distribution μ on F such that II.2.1 holds. Choose an arbitrary point x_0 and set

$$\phi(r) = \mu(\{x \mid |x - x_0| < r\}) .$$

Then, assuming as we may that $K \equiv 0$, $r > r_0$,

$$u_\mu(x_0) = \int K(r)\, d\phi(r) = -\int \phi(r)\, dK(r) \leq$$
$$\leq -\int h(r)\, dK(r) = \int K(r)\, dh(r) = \text{Constant} < \infty ,$$

since

$$K(\rho)\, h(\rho) = \int_0^\rho K(\rho)\, dh(r) \leq \int_0^\rho K(r)\, dh(r) \to 0$$

as $\rho \to 0$, and similarly for the ϕ-integral.

2. A convenient geometric criterion on vanishing capacity is expressed in the following theorem.

THEOREM 2. *If the set* E *can be covered by* $A(r)$ *closed spheres of radii* $\leq r$ *and*

$$-\int_0 \frac{K'(r)}{A(r)}\,dr = \infty, \tag{2.1}$$

then $C_K(E) = 0$.

Proof. We may clearly assume that all spheres have radii exactly $= r$ and that $A(r)$ denotes the lower bound. Let us assume that $C_K(E) > 0$. Then there is a mass distribution μ on E such that $I(\mu) < \infty$. We denote by $\mu(r, a)$ the mass distributed on $|x - a| \leq r$. We then find—the partial integrations can be shown to be justified as above and we assume $K \equiv 0$, $r > r_0$—

$$I(\mu) = \int d\mu(y) \int K(|x - y|)\, d\mu(x) = \int d\mu(y) \int_{r=0}^{\infty} K(r)\, d\mu(r, y) =$$

$$= \int d\mu(y) \left[-\int \mu(r, y)\, K'(r)\, dr\right] \geq \tag{2.2}$$

$$\geq \sum_{n=0}^{\infty} \int_{2^{-n-1}}^{2^{-n}} - K'(r)\, dr \int \mu(2^{-n-1}, y)\, d\mu(y).$$

Let us now assume that

$$E \subset \bigcup_{\nu=1}^{A_n} S_\nu^{(n)}, \quad A_n = A(2^{-n}),$$

where $S_\nu^{(n)}$ are closed spheres of radii 2^{-n}. Since A_n is minimal, it is easy to see that there is a constant C, only depending on the dimension d, such that every point $P \in E$ is contained in at most C spheres. The inequality (2.2) can be continued

$$I(\mu) \geq \sum_{n=0}^{\infty} \int_{2^{-n-1}}^{2^{-n}} - K'(r)\, dr\; C^{-1} \sum_{\nu=1}^{A_{n+2}} \int_{S_\nu^{(n+2)}} \mu(2^{-n-1}, y)\, d\mu(y) \geq$$

$$\geq C^{-1} \sum_{n=0}^{\infty} \int_{2^{-n-1}}^{2^{-n}} - K'(r)\, dr \sum_{\nu=1}^{A_{n+2}} \mu(S_\nu^{(n+2)})^2 .$$

By Schwarz's inequality we have

$$\mu(E)^2 \leq \Big(\sum_{\nu=1}^{A_n} \mu(S_\nu^{(n)})\Big)^2 \leq A_n \sum_{\nu=1}^{A_n} \mu(S_\nu^{(n)})^2 ,$$

whence

(2.3) $$\sum_{n=0}^{\infty} (K(2^{-n-1}) - K(2^{-n}))\, A_{n+2}^{-1} < \infty .$$

Since clearly

$$A_n \leq A_{n+1} \leq \text{Const.}\, A_n$$

and since $A(r)$ is a decreasing function of r, the integral (2.1) and the series (2.3) converge simultaneously. We have thus got a contradiction and so $C_K(E) = 0$.

3. *Some examples.* As tests of the results obtained we now consider certain sets of the Cantor type.

THEOREM 3. *Let E be the usual d-dimensional Cantor set so that the set E_n obtained in n:th step, consists of 2nd intervals with edges of length ℓ_n. Then E has positive K-capacity if and only if*

(3.1) $$\sum_{\nu} 2^{-\nu d} K(\ell_\nu) < \infty .$$

Proof. 1) Assume that (3.1) holds. Define μ_n of mass 1 with support on E_n and uniform density on E_n. Choose $x_0 \in E_n$. At a distance of $< \ell_{n-1}$ from x_0 there are at most $(4+1)^d$ intervals of E_n and at a distance $< \ell_{n-\nu}$ at most $(2 \cdot 2^\nu + 1)^d$ intervals. We find

$$\int K(|x_0 - y|)\, d\mu_n(y) \leq \text{Const.} \{\ell_n^{-d} \int_{|t|<\ell_n} 2^{-nd} K(|t|)\, dt +$$

$$+ \sum_{\nu=0}^{n} 2^{(\nu-n)d} K(\ell_{n-\nu})\} \leq \text{Const.}\,,$$

since

$$\ell_n^{-d} \int_{|t|<\ell_n} 2^{-nd} K(|t|)\, dt \leq \ell_n^{-d}\, 2^{-nd} \cdot \sum_{\nu=n}^{\infty} \int_{\ell_{\nu+1}<|t|<\ell_\nu} K(|t|)\, dt$$

$$\leq \text{Const.} \{K(\ell_{n+1})\, 2^{-nd} + K(\ell_{n+2}) \left(\frac{\ell_{n+1}}{\ell_n}\right)^d 2^{-nd} + \ldots\} \leq$$

$$\leq \text{Const.} \sum_{n}^{\infty} K(\ell_\nu)\, 2^{-\nu d} \to 0\,.$$

2) For the converse, we use Theorem 2 and observe that

$$A(\ell_n) \leq \text{Const.}\, 2^{nd}\,.$$

We find

$$-\int_0 \frac{K'(r)}{A(r)}\, dr = \sum^{\infty} - \int_{\ell_{n+1}}^{\ell_n} \frac{K'(r)}{A(r)}\, dr \geq$$

$$\geq \sum^{\infty} \frac{1}{A(\ell_{n+1})} (K(\ell_{n+1}) - K(\ell_n)) \geq$$

$$\geq \sum^{\infty} (K(\ell_{n+1}) - K(\ell_n)) \cdot \text{Const.}\, 2^{-nd} = \text{Const.} \sum^{\infty} K(\ell_n)\, 2^{-nd}\,.$$

Theorem 3 is now proved and we have shown that the geometric criterion in Theorem 2 is sharp. We shall now prove that the first condition of Theorem 1 also is best possible. In order to simplify the constructions we restrict ourselves to *linear* sets. The construction will be of the following generalized Cantor type.

Let $\{a_\nu\}$ and $\{b_\nu\}$ be non-increasing positive numbers such that $\Sigma\, b_\nu = \infty$. On $(0, 1)$ we construct *closed* I-intervals and *open* ω-intervals in the following way.

To the left we put $I_1^{(1)}$ of length a_1;

after $I_1^{(1)}$ we put $\omega_1^{(1)}$ of length b_1;

then follows $I_2^{(1)}$ of length a_2;

then follows $\omega_2^{(1)}$ of length b_2;

$\vdots$

then follows $I_{n_1}^{(1)}$ of length a_{n_1}.

We assume that the intervals constructed so far cover $(0, 1)$ exactly. The first approximation to the Cantor set is

$$E^{(1)} = \cup I_\nu^{(1)}$$

We next treat the interval $I_1^{(1)}$ in the same manner, starting with a_{n_1+1} resp. b_{n_1}. In this way we get I-intervals $I_\nu^{(2)}$, $\nu \leq n_\nu^{(2)}$ and corresponding ω-intervals. We continue with $I_2^{(1)}, \ldots, I_{n_1}^{(1)}$. Together we get in this second stage $I_\nu^{(2)}$, $1 \leq \nu \leq n_2$, and $\omega_\nu^{(2)}$, $1 \leq \nu \leq n_2 - n_1$, and we form

$$E^{(2)} = \cup I_\nu^{(2)}.$$

It is clear how the construction proceeds and we define

$$E = \bigcap_{n=1}^{\infty} E^{(n)}$$

The potentials, to be considered, are formed as follows. Let μ_1 be the unit mass distributed uniformly on the intervals $I_\nu^{(1)}$, the mass n_1^{-1} on every such interval. The mass on $I_\nu^{(1)}$, $1 \leq \nu \leq n_1$, is redistributed on its subinterval $I_\mu^{(2)}$, uniformly and with the same total mass on every subinterval of $I_\nu^{(1)}$. The resulting set function is called μ_2. This construction is repeated and we see that $\{\mu_n\}$ converges, $\mu_n \to \mu$, and that $S_\mu \subset E$. The potential corresponding to μ_n resp. μ and the kernel under consideration are called u_n resp. u.

We shall now prove the following theorem.

THEOREM 4. *For any $K(r)$ and any measure function $h(r)$ such that*

$$\varliminf_{r \to 0} h(r)\,\overline{K}(r) = 0$$

there is a set E as above with $C_K(E) > 0$ and $M_h(E) = 0$.

Proof. Define r_ν so that

$$h(r_\nu)\,\overline{K}(r_\nu) = \varepsilon_\nu^2 \to 0. \tag{3.4}$$

We shall determine a subsequence $\{r_{\nu_i}\}$ and choose the intervals $I_\nu^{(i)}$ of length r_{ν_i} and define

$$n_i = \left[\overline{K}(r_{\nu_i})\,\varepsilon_{\nu_i}^{-1}\right].$$

(3.4) then implies $M_h(E) = 0$.

Let us now consider the potentials $u_i(x)$ and assume that

$$u_i(x) \leq M_i.$$

When discussing an upper bound for $u_{i+1}(x)$, it is by the maximum principle sufficient to consider points $x_0 \in E^{(i+1)}$. Let $x_0 \in I_\nu^{(i)} = I$. We write

$$u_j'(x_0) = \int_I K(|x_0 - y|)\, d\mu_j(y).$$

$u_{i+1}(x_0)$ and $u_{i+1}'(x_0)$ depend on $r_{\nu_{i+1}}$ and $u_{i+1}(x_0) - u_{i+1}'(x_0)$ tends uniformly (with respect to x_0 and ν) to $u_i(x_0) - u_i'(x_0)$ as $r_{\nu_{i+1}} \to 0$. For $u_{i+1}'(x_0)$ we have the estimate, if we assume that I each I has been devided into m intervals $I_\mu^{(i+1)}$ at equal distance, and each observe how μ_{i+1} is obtained from μ_i:

$$\begin{aligned} u_{i+1}'(x_0) &\le 2m^{-1} \cdot \mu_i(I) \frac{1}{r_{\nu_{i+1}}} \int_0^{\frac{1}{2} r_{\nu_{i+1}}} K(t)\, dt + u_i'(x_0) \le \\ &\le \frac{\overline{K}(r_{\nu_{i+1}})}{m \cdot n_i} + u_i'(x_0) = \\ &= \frac{\overline{K}(r_{\nu_{i+1}})}{n_{i+1}} + u_i'(x_0) \le \\ &\le \varepsilon_{\nu_{i+1}} + u_i'(x_0). \end{aligned}$$

We can thus choose ν_{i+1} so large that for $x \in E^{(i+1)}$, e.g.,

$$u_{i+1}(x) \le u_i(x) + 2^{-i} \le M_i + 2^{-i}.$$

Hence the potentials $u_i(x)$ are uniformly bounded and so $C_K(E) > 0$.

The Cantor set shows that the second criterion of Theorem 1 cannot be improved. In the other direction, we shall now finally show that no complete description of capacity in terms of Hausdorff measure is possible.

THEOREM 5. *There is a set* E, *of the type used in Theorem 4, such that* $M_h(E) = 0$ *for every* $h(r)$ *such that*

$$\int_0 \frac{h(r)}{r}\, dr < \infty, \tag{3.5}$$

while the capacity is positive for $K(r) = \log \frac{1}{r}$.

Proof. Let us assume that $I_\nu^{(k)}$ have been constructed $1 \le k \le i$, $1 \le \nu \le n_i$, and that

$$u_k(x) \le M_k, \quad 1 \le k \le i .$$

We choose numbers $m_1, \dots, m_{n_i}$, $m_{j+1} > 2m_j$, sufficiently large. The $I^{(i+1)}$-intervals of $I_\nu^{(i)}$ are chosen of length

$$e^{-m_\nu}, e^{-m_\nu - 1}, \dots, e^{-2m_\nu}, \tag{3.6}$$

and the ω-intervals all of length e^{-q_ν}. Using the notations of the proof of Theorem 4 we see that, $x_0 \in I = I_\nu^{(i)}$,

$$u_{i+1}(x_0) - u'_{i+1}(x_0) \to u_i(x_0) - u'_i(x_0), \quad m_1 \to \infty ,$$

uniformly. Also

$$u'_{i+1}(x_0) \le \frac{1}{(m_\nu + 1)} \cdot \mu_i(I) \cdot 2e^{2m_\nu} \int_0^{\frac{1}{2}e^{-2m_\nu}} \log \frac{1}{t}\, dt + u'_i(x_0) \le$$

$$\le 2\mu_i(I) + u'_i(x_0) + O\left(\frac{1}{m_\nu}\right).$$

Choosing $m_1, \dots, m_{n_i}$ large enough we see as before that for given $x_0 \in E$, $\Sigma\, \mu_i(I)$ and $\Sigma \frac{1}{m_\mu}$ can be chosen uniformly bounded so that $C_K(E) > 0$, $K = \log \frac{1}{r}$. —We now observe that E can be covered by intervals of length (3.6), $\nu = 1, \dots, n_i$ and that every exponent occurs at most in one sequence (3.6). Hence

$$M_h(E) \le \sum_{j=m_1}^{\infty} h(e^{-j}) < \int_0^{e^{-m_1+1}} \frac{h(r)}{r} dr .$$

Hence $M_h(E) = 0$, if (3.5) holds.

4. The notion of capacity is connected with two elementary extremal problems, which historically were considered before capacities. —Let F be a compact set and define

$$(4.1) \qquad M_n = M_n(F) = \sup_{\{x_\nu\}} \inf_{x \in F} \frac{1}{n} \Sigma K(|x - x_\nu|)$$

and

$$(4.2) \qquad D_n = D_n(F) = \inf_{x_\nu \in F} \binom{n}{2}^{-1} \sum_{1 \leq \nu < \mu \leq n} K(|x_\nu - x_\mu|).$$

D_n is called generalized diameter (note $n = 2$) and M_n a Tscheby-scheff constant (on account of the significance of (4.1) for $d = 2$ and $K(r) = \log \frac{1}{r}$). The following theorem holds.

THEOREM 6. *If* $C_K(F) > 0$, *and* F *is compact, the limits* $\lim_{n \to \infty} M_n = M$ *and* $\lim_{n \to \infty} D_n = D$ *(transfinite diameter) exist and*

$$M^{-1} = D^{-1} = C_K(F).$$

If $C_K(F) = 0$, *then* $M_n \to \infty$ *and* $D_n \to \infty$.

Proof. We first assume $C_K(F) > 0$.

1) $D_{n+1} \leq M_n$.

Since $K(r)$ is continuous, the lower bound defining D_{n+1} is actually a minimum, i.e., $\{x_\nu\}_1^{n+1}$ exist so that

$$D_{n+1} = \frac{2}{n(n+1)} \sum_{i<j} K(|x_i - x_j|) = \frac{1}{n(n+1)} \sum_{i \neq j} K(|x_i - x_j|) =$$

$$= \frac{1}{n+1} \sum_{i=1}^{n+1} \frac{1}{n} \sum_{j \neq i} K(|x_i - x_j|).$$

If $x_1, x_2, \ldots, x_{i-1}, x_{i+1}, \ldots, x_{n+1}$ are considered fixed, x_i is chosen to minimize

$$a_i(x) = \frac{1}{n} \sum_{j \neq i} K(|x - x_j|)$$

for all $x \in F$. From the definition (4.1) we see that $a_i(x_i) \leq M_n$ and from the last expression for D_{n+1} above that

$$D_{n+1} = \frac{1}{n+1} \sum_{i=1}^{n+1} a_i(x_i) \leq M_n .$$

2) $M_n \leq C_K(F)^{-1} = V$.

By definition, there is a distribution μ of unit mass on F such that

$$\int K(|x - y|)\, d\mu(y) \leq V$$

everywhere. Hence

$$\inf_{x \in F} \frac{1}{n} \sum_{\nu=1}^{n} K(|x - x_\nu|) \leq \frac{1}{n} \sum_{\nu=1}^{n} \int_F K(|x - x_\nu|)\, d\mu(x) \leq V .$$

Taking the upper bound over $\{x_\nu\}$ we get $M \leq V$.

3) $V \leq \lim_{n \to \infty} D_n$.

Define μ_n to be the mass distribution with masses n^{-1} at those points x_ν which minimize D_n. Define $K^{(N)}(r)$:

$$K^{(N)}(r) = \mathrm{Min}\, (K(r), N) .$$

Then

$$I_{K^{(N)}}(\mu_n) \leq D_n + \frac{N}{n} .$$

Let $n \to \infty$ and choose n_ν so that $D_{n_\nu} \to \underline{\lim}\, D_{n_\nu}$ and so that $\mu_{n_\nu} \to \mu$. Then $S_\mu \subset F$ and

$$I_{K^{(N)}}(\mu) \leq \underline{\lim}\, D_n .$$

As $N \to \infty$ we obtain

$$V \leq I_K(\mu) \leq \underline{\lim}\, D_n \,.$$

Combining 1), 2) and 3) we now get

$$V \leq \underline{\lim}\, D_n \leq \left\{ \begin{matrix} \overline{\lim}\, D_n \\ \\ \underline{\lim}\, M_n \end{matrix} \right\} \leq \overline{\lim}\, M_n \leq V \,.$$

If $C(F) = 0$ choose $F' \supset F$ with $C(F') > 0$. Since

$$D_n(F) \geq D_n(F')$$

and

$$M_n(F) \geq M_n(F')$$

we find

$$\left. \begin{matrix} \underline{\lim}\, D_n(F) \\ \\ \underline{\lim}\, M_n(F) \end{matrix} \right\} \geq C_K(F')^{-1}$$

where the right hand side is arbitrarily large.

§V. *EXISTENCE OF BOUNDARY VALUES*

1. The most famous theorem on exceptional sets is Fatou's theorem on the existence of boundary values for a function bounded and analytic in the unit circle. This result has then been generalized in different directions, and a very general version will be given in section 2. In order to make the argument clear, an outline of a proof of Fatou's theorem is given here.

Let $u(z)$ be harmonic in $|z| < 1$ and assume $u(z) \geq 0$. By the Poisson formula we have for $r < R < 1$

$$(1.1) \quad u(re^{i\theta}) = \frac{1}{2\pi} \int_{-\pi}^{\pi} \frac{R^2 - r^2}{R^2 + r^2 - 2Rr\cos(\theta - \phi)} u(Re^{i\phi})\, d\phi .$$

In particular,

$$(1.2) \qquad u(0) = \frac{1}{2\pi} \int_{-\pi}^{\pi} u(Re^{i\phi})\, d\phi .$$

We can thus select a sequence $R_n \to 1$ so that $u(R_n e^{i\phi})\,d\phi$ converges weakly to some non-negative measure $d\mu$. We decompose $d\mu$ by Lebesgue's theorem:

$$(1.3) \qquad d\mu = f(\phi)\, d\phi + ds(\phi) ,$$

where $s(\phi)$ is singular. Formula (1.1) becomes

$$(1.4) \qquad u(re^{i\theta}) = \int_{-\pi}^{\pi} P(r; \theta - \phi)\left(f(\phi)\frac{d\phi}{2\pi} + \frac{1}{2\pi}\, ds(\phi)\right),$$

where P is the Poisson kernel, i.e. the normal derivative of the

Green's function.

The standard way to prove that

$$\lim u(z) \ \textit{exists a.e.,} \quad z \to e^{i\theta} \ \textit{non-tang.,} \tag{1.5}$$

is by means of a partial integration in (1.4) and Lebesgue's theorem on the existence of the derivative of an indefinite integral. This argument requires estimate of $\frac{\partial P}{\partial \theta}$, which makes generalizations difficult. However, using a slightly stronger version of Lebesgue's theorem we obtain a proof not depending on partial integrations and therefore possible to generalize.

It is well known that almost everywhere (θ)

$$\int_{-t}^{t} \{|f(\theta) - f(\theta+\phi)|\, d\phi + ds(\phi)\} = o(t), \quad t \to 0. \tag{1.6}$$

We assume that (1.6) holds for $\theta = 0$ and consider for simplicity only radial approach in (1.5). Choose $\delta > 0$, fixed as $r \to 1$, and define N so that $2^N\eta \leq \delta < 2^{N+1}\eta$, $\eta = 1 - r$. From (1.4) it follows

$$\begin{aligned} |u(r) - f(0)| &\leq \left\{\int_{-\eta}^{\eta} + \sum_{\nu=0}^{N} \int_{2^\nu\eta \leq |\phi| < 2^{\nu+1}\eta}\right\} |f(\phi) - f(0)|\, P\, \frac{d\phi + ds}{2\pi} + \\ &\quad + u(0) \operatorname*{Max}_{|\phi| \geq \delta} P(r,\phi) \leq \\ &\leq o(\eta) \operatorname{Max} P + \sum_{\nu=0}^{N} o(2^\nu\eta) \operatorname*{Max}_{2^\nu\eta \leq |\phi|} P + o(1) \leq \\ &\leq o\left\{\sum_{\nu=0}^{N} \frac{2^\nu}{2^{2\nu}}\right\} = o(1). \end{aligned} \tag{1.7}$$

2. We shall now use the above argument to prove a boundary value theorem for harmonic functions of several variables. The lack of the method of conformal mapping introduces technical difficulties in proofs of rather evident results. This fact is clearly illustrated in section 4.

Before stating the theorem we introduce some notations. We consider points $P = (x_1, x_2, \ldots, x_d; y) = (x; y)$ in $(d+1)$-dimensional Euclidean space. $|x|$ denotes distance on the d-dimensional subspace $X = \{P | y = 0\}$, dx denotes the volume element in X. By $V_\alpha(x^0)$ we mean the cone

$$V_\alpha(x^0) : |x - x^0| < \alpha y .$$

THEOREM 1. *Let* $u(P)$ *be harmonic in* $y > 0$ *and assume that for almost all* $x \in X$, *there is a cone* $V_\beta(x)$ *so that* $u(P)$ *is bounded from below in* $V_\beta(x)$. *Then*

$$\lim u(P), \; P \to (x; 0), \; P \in V_\alpha(x), \tag{2.1}$$

exists a.e. on X *for all* α.

Proof. We consider only x's belonging to some bounded set, e.g., $|x| < 1$. If we avoid an open subset O of measure $m\,O < \varepsilon$, we have for $y \leq y_0$ and a certain α independent of x, $u(P) \geq$ Const., $P \in V_\alpha(x)$, $x \notin O$. We form the region

$$R = R(O) = \{\bigcup_{x \notin O} V_\alpha(x)\} \cap \{P \,|\, |x| < 1, \; y < y_0\} .$$

If y_0 is large enough R is connected. We may assume that $u \geq 0$ in R. We observe that every boundary point P of R satisfies the Poincaré condition (some cone with vertex at P is contained in the complement of R). The Dirichlet problem can thus be solved for R. Let R_n be the part of R where $y > n^{-1}$ and let $G_n(P)$ be the Green's function for R_n with some fixed pole P_0. We need a uniform estimate of $G_n(P)$.

Let $\phi(t)$ denote the distance from $t \in O$ to the complement O' of O and form

$$h(x;y) = y \int_{O} \frac{\phi(t)\,dt}{\{(t-x)^2 + y^2\}^{(d+1)/2}} = y\,h_1(x;y).$$

$h(x;y)$ is harmonic in $y > 0$. Observing that $\phi(t) \geq \frac{1}{2}\phi(x)$ if $|t-x| \leq \frac{1}{2}\phi(x)$, we see that $h(x; C\phi(x)) \geq \lambda_d C^{-d}\phi(x)$, where λ_d only depends on d. (Points with $|x| = 1$ are also easily taken care of.) This implies that $h(x;z) \geq C'_a \cdot z$, $z = y - n^{-1}$, for $(x;y)$ on the part of ∂R_n, where $n^{-1} < y < y_0$, $|x| < 1$. Let $G_n^*(P)$ be the Green's function for the cylinder $n^{-1} < y < y_0$, $|x| < 1$, with pole at P_0. Clearly, if $\delta > 0$ is given, there exist two constants $c_1, c_2 > 0$ so that

$$c_1 z \leq G^*(P) \leq c_2 z$$

if $|x| < 1 - \delta$ and $y < \delta$ say. The second relation holds for all $|x| < 1$. By the maximum principle

$$G_n(P) \geq G^*(P) - C_a h(x;z) \quad \text{in} \quad R_n.$$

Hence for $c = c(\delta)$ independent of n and $|x| < 1 - \delta$, $y < \delta$,

$$G_n(P) \geq 2c(z - C_a h(x;z)) = 2c \cdot z(1 - C_a h_1).$$

We now need an estimate of $h_1(x;z) \leq h_1(x;0)$. We have

$$\int_{O'} h_1(x;0)\,dx \leq \int_{O} \phi(t)\,dt \int_{|x-t| \geq \phi(t)} \frac{dx}{|x-t|^{d+1}} \leq$$

$$\leq \lambda_d \int_{O} dt = \lambda_d mO < \lambda_d \varepsilon.$$

Hence $h_1(x;z) \leq (2C_a)^{-1}$ for all z, except when $x \in O_1$, $m\,O_1 < 2\lambda_d C_a \varepsilon = \varepsilon_1$.

What will be needed of the above investigation of G_n is that

$$\frac{\partial G_n}{\partial n} \geq c \quad \text{for all } n,\ P \in \partial R_n,\ y = n^{-1},$$

except for x in a set S of measure $< \varepsilon + \varepsilon_1$.

We now consider the harmonic measure $\omega_n(e;P)$ of a certain subset e of ∂R_n at a point $P \in R_n$. If $P = P_0$ we delete the variable P. Harnack's inequality yields

$$M(P)^{-1} \leq \frac{\omega_n(e;P)}{\omega_n(e)} \leq M(P)$$

with $M(P)$ independent of n and e, $n > n(P)$. We can write $d\omega_n(\cdot;P) = K_n(\cdot;P)\,d\omega_n$. Here $K_n(P)$ is harmonic in P and satisfies the inequality above. Also $K_n(P_0) = 1$. We form $u_\varepsilon(P) = u(x;y+\varepsilon)$ and have

$$u_\varepsilon(P) = \int_{\partial R_n} u_\varepsilon(Q)\,K_n(Q;P)\,d\omega_n(Q). \tag{2.2}$$

This formula corresponds to (1.1). Letting $n \to \infty$ we obtain with obvious notations

$$u_\varepsilon(P) = \int_{\partial R} u_\varepsilon(Q)\,K(Q;P)\,d\omega(Q).$$

Letting $\varepsilon \to 0$ we get for a certain $f \in L^1(d\omega)$ and with s singular with respect to ω

$$u(P) = \int_{\partial R} f(Q)\,K(Q;P)\,d\omega(Q) + \int_{\partial R} K(Q;P)\,ds(Q). \tag{2.3}$$

(2.3) is the analogue of (1.4).

3. Let us consider a point $Q_0 = (x;0) \in \partial R$ such that

(a) Q_0 is a point of density for the complement of S,

(b) $\displaystyle\int_{|x|<\varepsilon} |f(Q) - f(Q_0)|\,d\omega(Q) + \int_{|x|<\varepsilon} ds(Q) = o(\varepsilon^d),\ \varepsilon \to 0,\ Q = (x_0 + x;y).$

Since Lebesgue's theorem on symmetric derivatives holds for d dimensions, an inspection of the proof of (1.6) shows that (b) (as well as (a)) holds a.e. Namely, decompose $d\omega = \psi(Q)\,dQ + d\tau(Q)$ where $\psi \in L^1(dQ)$ and τ is singular with respect to Lebesgue measure. Then $f \in L^1(d\tau)$ and

$$\int_{|x|<\varepsilon} |f(Q) - f(Q_0)|\,\psi dQ \leq \int_{|x|<\varepsilon} |f(Q)\,\psi(Q) - f(Q_0)\,\psi(Q_0)|\,dQ + \\ + \int_{|x|<\varepsilon} f(Q_0)\,|\,\psi(Q) - \psi(Q_0)|\,dQ = o(\varepsilon^d)$$

almost everywhere. Since τ is singular

$$\int_{|x|<\varepsilon} f(Q)\,d\tau(Q) = o(\varepsilon^d) \quad \text{a.e.}$$

We finally observe that $\frac{\partial G_n}{\partial n} \geq c$ for $(x;y) \in \partial R_n$, $x \notin S$. Since the surface element $d\sigma_n$ of ∂R_n also satisfies an inequality $d\sigma_n > c\,dQ$, it follows that s is singular also with respect to Lebesgue measure.

Let us assume that $Q_0 = (0;0)$ is a point, where (a) and (b) hold. We choose $A = (0;a)$, $a > 0$, and consider $u(A)$, as $a \to 0$. The general non-tangential approach is analogous. Define for a fixed $\delta > 0$

$$K_\nu = \{Q \mid Q \in \partial R,\ y < y_0,\ |x_i| < 2^\nu a\}$$

for $\nu = 0, 1, \ldots, N$, $2^N a \leq \delta < 2^{N+1} a$, and

$$L_\nu = K_\nu - K_{\nu-1}, \quad \nu = 1, \ldots, N, \quad L_0 = K_0,$$

and

$$\Gamma = \partial R - K_N .$$

Formula (2.3) yields (cf. (1.7))

$$|u(A)-f(Q_0)| \le |\int (f(Q)-f(Q_0))K(Q;A)\,d\omega(Q)| + \int K(Q;A)\,ds(Q) \le$$

$$\le \sum_{\nu=0}^{N} \sup_{Q\in L_\nu} K(Q;A)\varepsilon(\delta)\, 2^{d\nu} a^d + O(1) \sup_{Q\in\Gamma} K(Q;A).$$

We must study the harmonic functions $K(Q;A)$ for $Q = Q^{(\nu)} \in L_\nu$ and consider first the case $\nu = 0$.

Since $\partial G_n/\partial n \ge c$ for $(x;y) \in \partial R_n$, $x \notin S$, it follows from condition (a) that the harmonic measure $v_0(P)$ of L_0 satisfies

$$v_0(P_0) \ge \gamma\, a^d, \tag{3.2}$$

where the constant γ is independent of a. We also observe that ∂R, $|x| < 1$, can be represented $y = \psi(x)$, where ψ satisfies a Lipschitz condition of order 1 and $\psi(x) = o(|x|)$, $|x| \to 0$.

We remove from R the set $|x_i| < 2a$, $y < ka$. The resulting domain is called R'. The harmonic measure of the part of $\partial R'$ with $|x_i| < 2a$ is called $v_0'(P)$. Since the harmonic measure of $\{P \mid P \in \partial R',\ |x_i| = 2a,\ y < ka\}$ with respect to R' is smaller than the harmonic measure of the same set with respect to $y > 0$, it follows that its value at $P_0 = O(k)a^d$. Hence $v_0'(P)$ also satisfies that inequality (3.2) if k is small enough.

We set $K(Q^{(0)}; A) = \mu_0$. From Harnack's inequality and the maximum principle it follows that

$$K(Q^{(0)}; P) \ge \text{Const.}\ \mu_0\, v_0'(P).$$

Setting $P = P_0$ we find

$$\mu_0 \le \text{Const.}\ a^{-d}. \tag{3.3}$$

We now choose $Q = Q^{(\nu)} = (x_\nu; y_\nu)$ and consider $B = (x_\nu; 2^\nu a)$, $\nu \le N$. By (a) $\psi(x_\nu) = o(2^\nu a)$. We set $K(Q^{(\nu)}; B) = \mu_\nu$ and find as above

$$\mu_\nu \leq \text{Const. } a^{-d} 2^{-d\nu} .$$

On the other hand, $K(Q^{(\nu)}; P)/\mu_\nu$ is a positive harmonic function which vanishes on $\partial R - L_\nu$ and $= 1$ for $P = B$. (In fact, one should first consider K_n; since all estimates are uniform, $n \to \infty$ causes no difficulty.) By Lemma 1 in section 4 and the maximum principle

$$K(Q^{(\nu)}; P) \leq \text{Const. } \mu_\nu \int_{(t;\, \psi(t) \in K_{\nu+1} - K_{\nu-2}} \frac{y\, dt}{\{(x-t)^2 + y^2\}^{(d+1)/2}}$$

for $P \in R$, $|P - Q^{(\nu)}| \geq \text{Const.} \cdot 2^\nu a$. Inserting $P = A$ we find

$$K(Q^{(\nu)}; A) \leq \text{Const. } 2^{-\nu}\, 2^{-\nu d}\, a^{-d} . \tag{3.4}$$

Finally, if $Q \in \Gamma$, the argument giving (3.4) can be used for $\nu = N$ giving $\sup_{Q \in \Gamma} K(Q; A) \to 0$, $a \to 0$. Inserting (3.4) in (3.1) we find $\lim_{a \to 0} u(A) = f(Q_0)$ and the theorem is proved.

4. LEMMA 1. *Let E be a subset of X in $|x| < 1$ and form for a fixed α*

$$R = \bigcup_{x \in E} V_\alpha(x) \cap \{P \mid |x| < 1,\, y < 1\}$$

and assume that the part Γ of ∂R with $|x| < 1$, $y < 1$ satisfies $y < \frac{1}{3}$. Let u be a positive harmonic function in R which vanishes continuously on ∂R except on the part of Γ which satisfies $|x| < \frac{1}{3}$. Then there exists a constant K, only depending on α, such that

$$u(x; y) \leq K \cdot u(0; \tfrac{1}{2}), \quad |x| = \frac{1}{2} . \tag{4.1}$$

Proof. By (2.2) it is sufficient to prove (4.1) when u is the harmonic measure of $\Gamma \cap \{P \mid |x - x_0| < \rho\}$ for ρ arbitrarily small and

$|x_0| < \frac{1}{3}$. To simplify the notations we choose $x_0 = 0$. The proof shows that this is no restriction. We use the notation K_i for constants only depending on α.

Suppose that $(0; y_0) \in \Gamma$ and consider the sets D_ν:

$$D_\nu = R \cap \{P \,\big|\, |x| < 2^\nu \rho,\ y \leq y_0 + K_1 2^\nu \rho = \eta_\nu\},\quad \nu = 0, 1, \ldots .$$

If K_1 is large enough the boundary of D_ν consists of three parts:

(1) a subset α_ν of Γ;

(2) a subset β_ν of the cylinder $|x| = 2^\nu \rho$;

(3) a "circle" γ_ν: $|x| < 2^\nu \rho$, $y = \eta_\nu$.

We use the notation $q_\nu = u(0; \eta_\nu)$. If K_1 is large enough it follows from Harnack's principle that

$$u(P) \leq K_2\, q_\nu \text{ on } \gamma_\nu \tag{4.2}$$

and

$$q_{\nu-1} \leq K_2\, q_\nu . \tag{4.3}$$

To be able to discuss $u(P)$ on β_ν we observe that R has the following property. If ξ is a given x-vector such that $|\xi| = 2^\nu \rho$ and $\eta(\xi) < y < \eta_\nu$ is the corresponding subset of β_ν, then $\eta_\nu - \eta(\xi) <$ $< K_3 2^\nu \rho$ and all points $(x; y)$ with $|x - \xi| < \alpha(y - \eta(\xi))$, $|x| < 1$, $y < 1$, belong to R. δ is a positive number to be determined later and we write $K_i(\delta)$ for functions of α and δ. (4.2) and the above mentioned property of R imply, again by Harnack's inequality, that

$$u(\xi; y) \leq K_4(\delta)\, q_\nu,\quad \eta(\xi) + \delta\, 2^\nu \rho < y < \eta_\nu . \tag{4.4}$$

We shall now show by induction that, for ρ small enough,

$$u(P) \leq K_5\, q_j,\quad P \in \beta_j \cup \gamma_j . \tag{4.5}$$

Let us first consider $j = 0$. That (4.5) holds in this case is easily seen if we compare u with the harmonic measure of the bottom of a cylinder with radius ρ and side $K_6 \rho$, evaluated at its center of gravity. We now assume that (4.5) holds for $j \leq \nu - 1$. To prove (4.5) on $\beta_\nu \cup \gamma_\nu$ it is, by (4.2) and (4.4) only the part of β_ν with $\eta(\xi) < y \leq \eta(\xi) + \delta\, 2^\nu \rho$ that has to be considered.

Let Σ be the following auxiliary domain

$$\Sigma : P \in \Sigma \text{ if, } ay > -|x|,\ |x - (-1, 0, \ldots, 0)| > \frac{1}{2}$$

and let $\omega(P)$ be the harmonic measure of the part of $\partial\Sigma$ which is not the cone $ay = -|x|$.

We now shrink Σ by a length factor $2^{\nu+1}\rho$ and make a translation and rotation of the resulting domain to a domain with vertex of the cone at $(\xi; \eta(\xi))$ and cylinder axis along the y-axis. ω becomes ω_1 and it follows from the maximum principle, the induction assumption and (4.3) that

$$u(P) \leq K_5\, q_{\nu-1}\, \omega_1(P) \leq K_7\, q_\nu\, \omega_1(P)$$

in $D_\nu - D_{\nu-1}$. Since $\omega(0; y) \to 0$, $y \to 0$, it follows that

$$\omega_1(\xi; \eta(\xi) + s\, 2^\nu \rho) < \varepsilon$$

if $s < \delta(\varepsilon)$. Hence if $\delta = \delta(K_7^{-1} K_5)$, (4.5) is proved for $j = \nu$.

The induction can be continued as long as $2^\nu \rho \leq \frac{1}{2}$. The maximum principle now shows that (4.1) holds.

5. Theorem 1 is easily seen to be best possible as to the size of the exceptional set. Let E be measurable on the unit circle $(d = 2)$ of measure zero. Choose open sets $O_n \supset E$, $m\, O_n < 2^{-n}$, and define

$$u(re^{i\theta}) = \sum_{n=0}^{\infty} \frac{1}{2\pi} \int_{O_n} \frac{1 - r^2}{1 + r^2 - 2r\cos(\theta - \phi)}\, d\phi\, .$$

Then $u(0) = \Sigma\, 2^{-n}$ so u is positive and harmonic in $r < 1$ and for $\theta \in E$

$$\underline{\lim}_{r \to 1}\, u(re^{i\theta}) \geq \sum_{n=0}^{N} 1 \geq N + 1 \text{ for all } N .$$

We shall now consider restrictions on u which guarantee a smaller exceptional set. It then turns out that, on one hand, the assertion on the existence of boundary values for *harmonic* functions can be strengthened to convergence of the corresponding Fourier series (the well-known Kolmogoroff example shows that this fails for the general Fatou theorem), while, on the other hand, the existence of boundary values holds for a general class of functions. In this section we prove the first assertion.

THEOREM 2. *Let* $K(r)$ $(d = 1)$ *have the properties* $K(0) = \infty$ *and* $K(r) \equiv 0$, $r > 1$, *and define*

$$\lambda_n^{-1} = \int_0^1 K(x) \cos nx \, dx . \tag{5.1}$$

Then, by Lemma III.2, $\lambda_n > 0$. *Consider the Fourier series*

$$\sum_1^\infty (a_n \cos nx + b_n \sin nx) \tag{5.2}$$

and assume

$$\sum_1^\infty (a_n^2 + b_n^2)\, \lambda_n < \infty . \tag{5.3}$$

Then (5.2) *converges except when* x *belongs to a set* E *such that* $C_{\overline{K}}(E) = 0$, *where, as before*

$$\overline{K}(r) = \frac{1}{r} \int_0^r K(x)\, dx .$$

Conversely, if E *is closed and* $C_K(E) = 0$ *and* $K(x)$ *satisfies* $K(x) = O(K(2x))$, $x \to 0$, *there is a series* (5.2), *satisfying* (5.3), *which diverges on* E.

The proof depends on the following lemma.

LEMMA 2. *With the notations of Theorem 2, there is a constant* M *so that for all* x *and* n, $-\pi \leq x \leq \pi$,

$$\left| \sum_{\nu=1}^{n} \frac{\cos \nu x}{\lambda_\nu} \right| \leq M\, \overline{K}(|x|).$$

Proof. Define

$$k_n(x) = \frac{1}{2}\lambda_0^{-1} + \sum_{\nu=1}^{n} \frac{\cos \nu x}{\lambda_\nu} = \int_0^1 \left(\frac{1}{2} + \sum_{\nu=1}^{n} \cos \nu x \cos \nu t\right) K(t)\, dt =$$

$$= \int_0^1 (D_n(x+t) + D_n(x-t)) K(t)\, dt = I_1 + I_2$$

where $D_n(x)$ is the Dirichlet kernel

$$D_n(x) = \frac{\sin (n + \frac{1}{2})x}{4 \sin \frac{x}{2}} .$$

$D_n(x)$ satisfies the inequalities

$$|D_n(x)| < \frac{1}{4\,|\sin \frac{x}{2}|} < \frac{M_1}{|x|} \tag{5.4}$$

and

$$\left| \int_0^x D_n(t)\, dt \right| \leq M_2 . \tag{5.5}$$

We find, if $x > 0$,

$$|I_1| = \left| \int_0^1 D_n(x+t) K(t)\, dt \right| \leq \left| \int_0^x \right| + \left| \int_x^1 \right| = J_1 + J_2 .$$

Here, by the inequality (5.4),

$$J_1 \le \frac{M_1}{x} \int_0^x K(t)\,dt = M_1 \bar{K}(x)$$

and by (5.5)

$$J_2 = \left| \int_x^1 \left(\int_x^t D_n(u)\,du \right) \right| K'(t)|\,dt \le 2\,M_2\,K(x) \le 2\,M_2\bar{K}(x).$$

I_2 is decomposed

$$I_2 = \left| \int_0^{x/2} \right| + \left| \int_{x/2}^{3x/2} \right| + \left| \int_{3x/2}^{1} \right.$$

and these terms are estimated as above.

Proof of Theorem 2. 1) We assume (5.2) given, satisfying (5.3) and form

$$q_n(x) = \sum_{\nu=1}^{n} \frac{\cos \nu x}{\sqrt{\lambda_\nu}} .$$

There are functions $F_p(x) \in L^2(-\pi, \pi)$ with Fourier series

$$F_p(x) \sim \sum_{\nu=p}^{\infty} (a_\nu \cos \nu x + b_\nu \sin \nu x) \sqrt{\lambda_\nu} .$$

The partial sums of the given series can be written

$$s_n(x) = \frac{1}{\pi} \int_{-\pi}^{\pi} F_1(t)\, q_n(x-t)\,dt .$$

Let $n(x)$ be an arbitrary Borel-measurable function, $n(x) \le N$, taking integer values and let μ be a distribution of mass on $(-\pi, \pi)$. Then

$$\left(\int_{-\pi}^{\pi} s_{n(x)}(x)\,d\mu(x) \right)^2 = \left[\frac{1}{\pi} \int_{-\pi}^{\pi} F_1(t) \left(\int_{-\pi}^{\pi} q_{n(x)}(x-t)\,d\mu(x) \right) dt \right]^2 \le$$

$$\leq \frac{1}{\pi} \int_{-\pi}^{\pi} F_1(t)^2 dt \iint_{-\pi}^{\pi} d\mu(x) d\mu(y) \frac{1}{\pi} \int_{-\pi}^{\pi} q_{n(x)}(x-t) \cdot q_{n(y)}(y-t) dt =$$

$$= \|F_1\|^2 \iint_{-\pi}^{\pi} k_{n(x,y)}(x-y)\, d\mu(x)\, d\mu(y) ,$$

where $n(x, y) = \text{Min}\,(n(x), n(y))$. Lemma 2 yields

$$\left(\int_{-\pi}^{\pi} s_{n(x)}(x)\, d\mu(x)\right)^2 \leq M \;\|F_1\|^2 \iint \overline{K}(|x-y|)\, d\mu(x)\, d\mu(y). \tag{5.6}$$

Let A be a closed set where $\overline{\lim}\, s_n(x) \geq a$. Choosing μ to be the equilibrium distribution of A of mass 1 and $n(x)$ appropriately, (5.6) yields

$$C_{\overline{K}}(A) \leq M \;\|F_1\|^2 a^{-2} . \tag{5.7}$$

The set E where $\overline{\lim}\, s_n(x) - \underline{\lim}\, s_n(x) \geq a > 0$ does not depend on the first p Fourier coefficients. We can thus, if $A \subset E$, replace $\|F_1\|$ by any $\|F_p\|$ and (5.7) yields $C_{\overline{K}}(A) = 0$. This proves Theorem 2,1).

2) Assume E closed with $C_K(E) = 0$. Then a finite sum of intervals F with $C_K(F) = \varepsilon$ exists for any $\varepsilon > 0$ and a distribution of mass ε on F such that

$$u(x) = \int_F K(|x-y|)\, d\mu(y) \equiv 1$$

on F and

$$I(\mu) = \iint K(|x-y|)\, d\mu(x)\, d\mu(y) = \varepsilon ,$$

while $E \subset$ interior of F. We assume $E \subset (0, 1)$ and define μ and u periodically (2π). If disregarding $\nu = 0$,

$$d\mu \sim \Sigma\, (a_\nu \cos \nu x + \beta_\nu \sin \nu x)$$

then

$$u(x) \sim \Sigma\,(\alpha_\nu \cos \nu x + \beta_\nu \sin \nu x)\,\lambda_\nu^{-1}$$
$$\sim \Sigma\,(a_\nu \cos \nu x + b_\nu \sin \nu x)$$

and

$$I(\mu) = \Sigma\,(\alpha_\nu^2 + \beta_\nu^2)\,\lambda_\nu^{-1} = \Sigma\,(a_\nu^2 + b_\nu^2)\,\lambda_\nu .$$

We now choose $F^{(1)} \supset F^{(2)} \supset \ldots \supset E$ so that $C_K(F^{(n)}) = \varepsilon_n < 2^{-n}$. Corresponding u's and μ's are also denoted with upper indices. For a sequence of integers $\{n_i\}$ to be determined later we define

$$u(x) = \Sigma\, u^{(n_i)}(x) \sim \Sigma\,(a_\nu \cos \nu x + b_\nu \sin \nu x)$$

and

$$\mu = \mu^{(n_i)} .$$

We have

$$I(\mu) < 2 \sum_{i \leq j} I(\mu^{(i)}, \mu^{(j)}) \leq 2 \sum_1^\infty i\, \varepsilon_i < 4 .$$

We have $\sum_{i=1}^{j} u^{(n_i)}(x) \equiv j$ on a neighbourhood of E. We assume that for the partial sums $s_k^{(j)}(x)$ of the Fourier series of $\sum_1^j u^{(n_i)}(x)$, it is true that $s_k^{(j)}(x) > j-1$ for $k \geq N_j$ and $x \in F^{(n_j)}$. If we now choose n_{j+1} large, $u^{(n_{j+1})}(x)$ influences the Fourier coefficients of $\sum_1^{j+1} u^{(n_i)}(x)$ of order $\leq N_j$ arbitrarily little, while on the other hand we have $s_k^{(j+1)}(x) > j$ for $k \geq N_{j+1}$ and $x \in F^{(n_{j+1})}$. If the sequence $\{n_i\}$ is determined successively in this manner, the function u is an example of the desired kind.

6. We shall now restrict ourselves to capacities with respect to $r^{-\alpha}$

and $\log \frac{1}{r}$ and use the notations $C_{\alpha}(E)$ and $C_0(E)$. The following general theorem on existence of radial limits will be proved.

THEOREM 3. *Let $f(z)$ be continuous in $|z| < 1$, $z = x + iy$, having first partial derivatives a.e. in $|z| < 1$ such that*

$$\iint_{|x|<1} |\operatorname{grad} f|^2 (1-|z|)^{\alpha}\, dx\, dy < \infty, \quad 0 \le \alpha < 1, \tag{6.1}$$

and of class BL, i.e., such that $f(re^{i\theta})$ is absolutely continuous for almost all r as a function of θ and for almost all θ as a function of r. Then

$$\lim_{r \to 1} f(re^{i\theta}) = f(\theta)$$

exists except on a set E, $C_{\alpha}(E) = 0$.

For the proof we need some lemmas.

LEMMA 3. *Let $q(\zeta)$ be Borel-measurable on $|\zeta| < 1$ and assume*

$$\|q\|^2 = \iint_{|\zeta|<1} |q(\zeta)|^2 d\xi\, d\eta < \infty, \quad \zeta = \xi + i\eta.$$

Let A be the subset of $|z| < 1$ where

$$Q(z) = \iint_{|\zeta|<1} \frac{|q(\zeta)|}{|z-\zeta|}\, d\xi\, d\eta > a.$$

Then for a fixed constant C,

$$C_0(A) < C\,\frac{\|q\|^2}{a^2}. \tag{6.2}$$

Proof. Let μ be a distribution of unit mass on a subset A_1 of A of diameter $\le \frac{1}{3}$.

$$a^2 < \left(\int_{A_1} Q(z)\, d\mu(z) \right)^2 \le \|q\|^2 \iint_{A_1 A_1} d\mu(z_1)\, d\mu(z_2) \iint_{|\zeta| < 1} \frac{d\xi\, d\eta}{|z_1 - \zeta|\, |z_2 - \zeta|}.$$

It is easy to see that

(6.3)
$$\iint_{|\zeta| < 1} \frac{d\xi\, d\eta}{|z_1 - \zeta|\, |z_2 - \zeta|} < C_1 \log \frac{1}{|z_1 - z_2|},$$

which yields

(6.4)
$$C_0(A_1) < C_1 \frac{\|q\|^2}{a^2}.$$

Adding a bounded number of inequalities (6.4), we get (6.2).

LEMMA 4.   *Let* $\zeta_\nu = \rho_\nu e^{i\theta_\nu}$, $\nu = 1, 2$, *satisfy* $\frac{1}{2} < \rho_\nu < 1$. *Then for a fixed constant* $C = C_\alpha$, *depending only on* α,

(6.5)
$$\iint_{|z| < 1} \frac{dx\, dy}{(1 - |z|)^\alpha |z - \zeta_1|\, |z - \zeta_2|} \le C \begin{cases} \left| e^{i\theta_1} - e^{i\theta_2} \right|^{-\alpha}, & 0 < \alpha < 1 \\[2mm] \log \dfrac{3}{\left| e^{i\theta_1} - e^{i\theta_2} \right|}, & \alpha = 0. \end{cases}$$

Proof.   The inequality for $\alpha = 0$ is equivalent to (6.3) and we shall in the proof assume $0 < \alpha < 1$.

Mapping $|z| < 1$ conformally (or even only with bounded dilatation in both directions in the considered range) onto a half plane, we see that we may replace (6.5) by

$$H(\zeta_1, \zeta_2) = \int_0^\infty \int_{-\infty}^\infty \frac{dx\, dy}{y^\alpha |z - \zeta_1|\, |z - \zeta_2|}, \quad z = x + iy,$$

where $\zeta_\nu = \xi_\nu + i\eta_\nu$ and $0 < \eta_1 \le \eta_2$. H is then a decreasing function of η_2, $\eta_1 \le \eta_2 < \infty$. We can thus assume $\eta_1 = \eta_2 = \eta$. For ξ_1 and ξ_2 fixed, H is a decreasing function of $\eta > 0$, whence we can assume $\eta_1 = \eta_2 = 0$. H then depends on $|\xi_1 - \xi_2|$ only and we put

$\xi_2 = 0$ and $\xi_1 = \xi > 0$. We now have

$$H = \int_0^\infty \frac{dy}{y^a} \int_{-\infty}^0 \frac{dx}{|z|\,|z-\xi|} = \int_0^\infty \frac{dy}{y^a} I(y,\xi).$$

In the integral I we distinguish two cases.

1) $y \geq \xi$.

$$I(y,\xi) \leq 2\int_{-\infty}^{\infty} \frac{dx}{|z|^2} + \int_0^{2\xi} \frac{dx}{y^2} = \frac{\pi}{y} + \frac{2\xi}{y^2}.$$

2) $y \leq \xi$.

$$I(y,\xi) \leq 2\int_{-\infty}^{-\xi} \frac{dx}{x^2} + 2\int_{-\xi}^{\xi/2} \frac{dx}{\sqrt{x^2+y^2}} \cdot \frac{2}{\xi} <$$

$$< \frac{2}{\xi} + \frac{8\sqrt{2}}{\xi} \int_0^{\xi} \frac{dx}{x+y} = \frac{2}{\xi} + \frac{8\sqrt{2}}{\xi} \log \frac{\xi + y}{y}.$$

The contribution H_1 to H from the estimate 1) is majorized

$$H_1 < 2\pi \int_\xi^\infty \frac{dy}{y^{1+a}} + \xi \int_\xi^\infty \frac{dy}{y^{2+a}} = C'\,\xi^{-a}$$

and the analogous quantity H_2 satisfies the inequality

$$H_2 < \frac{2}{\xi}\int_0^\xi \frac{dy}{y^a} + \frac{8\sqrt{2}}{\xi}\int_\xi^\infty \frac{1}{y^a} \log\frac{y+\xi}{y}\,dy < C''\,\xi^{-a}$$

as a partial integration shows.

LEMMA 5. *Suppose that* $a(t)$, $0 \leq t \leq 1$, *satisfies the conditions*

$$a(t) \text{ absolutely continuous}, \quad a'(t) \in L^2(0,1), \quad a(0) = 0.$$

Then for $0 \leq a < 1$, *there is a constant* C *only depending on* a, *such that for any* $n \geq 1$

$$(6.6) \qquad \int_0^1 (|a'(t)|^2 + n^2|a(t)|^2)(1-t)^{\alpha}dt \geq C\, n^{1-\alpha}|a(1)|^2 .$$

Proof. Let τ be the largest number such that

$$|a(\tau)| = \frac{1}{2}|a(1)| .$$

Such a number clearly exists since $a(0) = 0$ and $\tau < 1$, except in the case when $a(1) = 0$. We find

$$\int_0^1 (|a'|^2 + n^2|a|^2)(1-t)^{\alpha}dt \geq \int_\tau^1 |a'|^2(1-t)^{\alpha}dt + \frac{1}{4}n^2|a(1)|^2\int_\tau^1 (1-t)^{\alpha}dt \geq$$

$$\geq \left(\int_\tau^1 |a'|dt\right)^2\left(\int_\tau^1 (1-t)^{-\alpha}dt\right)^{-1} + \frac{1}{4}(1+\alpha)^{-1}n^2|a(1)|^2\cdot(1-t)^{1+\alpha} \geq$$

$$\geq \frac{|a(1)|^2}{4}\left\{\frac{1-\alpha}{(1-\tau)^{1-\alpha}} + \frac{n^2}{1+\alpha}(1-\tau)^{1+\alpha}\right\}.$$

Minimizing the last expression as a function of $(1-\tau)$ we obtain (6.6).

Proof of Theorem 3. We first assume that f_x and f_y are continuous in $|z| \leq \rho < 1$. We apply Green's formula to $f(z)$ and $(z-\zeta)^{-1}$ in $|z| \leq \rho$, $|z-\zeta| \geq \varepsilon$ for a fixed ζ in $|\zeta| < \rho$ and let $\varepsilon \to 0$. We get, using the notation $f_{\bar z} = \frac{1}{2}(f_x + if_y)$,

$$(6.7) \qquad f(\zeta) = \frac{1}{2\pi i}\int_{|z|=\rho} \frac{f(z)}{z-\zeta}\,dz + \frac{1}{\pi}\iint_{|z|\leq\rho} \frac{f_{\bar z}}{z-\zeta}\,dx\,dy .$$

To obtain (6.7) in the general case, choose f_n with continuous first derivatives so that

$$\iint_{|z|\leq\rho} |\operatorname{grad}(f_n - f)|^2 dx\,dy = \varepsilon_n \to 0, \quad n \to \infty,$$

and so that $f_n(z) \to f(z)$ uniformly in $|z| \leq \rho$. (6.7) holds for f_n and the left hand side and the first term to the right clearly tend to the corresponding expressions for f, uniformly in $|\zeta| \leq \rho' < \rho$. Let A_n be the set in $|\zeta| \leq \rho$ where

$$\Bigl| \iint_{|z| \leq \rho} \frac{f_{\bar{z}}}{z-\zeta}\, dx\, dy - \iint_{|z| \leq \rho} \frac{f_{n\bar{z}}}{z-\zeta}\, dx\, dy \Bigr| > a_n .$$

By Lemma 3 we have

$$C_0(A_n) < C\, \varepsilon_n a_n^{-2} .$$

Choosing $\varepsilon_n \to 0$, $a_n \to 0$ so that $\Sigma\, \varepsilon_n a_n^{-2} < \infty$ we see that (6.7) *holds for the given function* f *except for* ζ *on a set* E_ρ *such that* $C_0(E_\rho) = 0$.

We now study (6.7) as $\rho \to 1$ and expand the first term to the right:

$$\frac{1}{2\pi i} \int_{|z| = \rho} \frac{f(z)}{z-\zeta}\, dz = \frac{1}{2\pi} \int_{-\pi}^{\pi} \frac{f(\rho e^{i\theta})}{1 - \zeta z^{-1}}\, d\theta = [z = \rho e^{i\theta}]$$

$$(6.8) \qquad = \sum_{n=0}^{\infty} \zeta^n \rho^{-n} \frac{1}{2\pi} \int_{-\pi}^{\pi} f(\rho e^{i\theta}) e^{-in\theta} d\theta =$$

$$= \sum_{n=0}^{\infty} \zeta^n \rho^{-n} a_n(\rho) ,$$

where

$$a_n(\rho) = \frac{1}{2\pi} \int_{-\pi}^{\pi} f(\rho e^{-i\theta}) e^{-in\theta} d\theta , \quad n = 0, \pm 1, \pm 2, \ldots .$$

This yields formally

$$a_n'(\rho) = \frac{1}{2\pi} \int_{-\pi}^{\pi} f_\rho(\rho e^{i\theta}) e^{-in\theta} d\theta$$

and

$$in\, a_n(\rho) = \frac{1}{2\pi} \int_{-\pi}^{\pi} f_\theta(\rho e^{i\theta}) e^{-in\theta} d\theta ,$$

which then by assumption on absolute continuity holds for almost all ρ. By (6.1) and Parseval's relation we have

$$\sum_{n=-\infty}^{\infty}\int_0^1 (|a_n'(\rho)|^2 + n^2|a_n(\rho)|^2)(1-\rho)^\alpha d\rho < \infty,$$

if we e.g. assume $f \equiv 0$ for $|z| < \frac{1}{2}$. By Lemma 5 there is a constant C so that for any $R < 1$

$$\sum_{n=0}^{\infty} |a_n(R)|^2 n^{1-\alpha} \leq C.$$

Choosing $R_\nu \nearrow 1$ so that $\lim_{\nu\to\infty} a_n(R_\nu) = a_n$ exist for $n \geq 0$ we find

(6.9) $$\sum_{n=0}^{\infty} |a_n|^2 n^{1-\alpha} \leq C$$

and

(6.10) $$g(\zeta) = \lim_{\nu\to\infty} \frac{1}{2\pi i} \int_{|z|=R_\nu} \frac{f(z)}{z-\zeta} dz = \sum_{n=0}^{\infty} a_n \zeta^n, \quad |\zeta| < 1.$$

We now choose $\rho = R_\nu$ in (6.7) and assume $\zeta \notin E$,

$$E = \bigcup_{\nu=1}^{\infty} E_{R_\nu},$$

so that $C_0(E) = 0$. Then

(6.11) $$f(\zeta) = g(\zeta) + \frac{1}{\pi} \iint_{|z|<1} \frac{f_{\bar{z}}}{z-\zeta} dx\, dy, \quad \zeta \notin E.$$

By Theorem 2, $g(\zeta)$ has radial limits except on a set of α-capacity zero. The exceptional set E where (6.11) does not hold is situated on radii corresponding to arguments belonging to a set of vanishing 0-capacity. Since such a set a fortiori has vanishing α-capacity, it is now sufficient to prove that the second term of (6.11) has radial limits except on a set of α-capacity zero.

Assume then that these limits do not exist on G, $C_\alpha(G) > 0$. To any choice of $R < 1$, $\rho(\theta)$ must exist so that for $\theta \in G_R \subset G$, $C_\alpha(G_R) \geq \frac{1}{2} C_\alpha(G)$, $\rho(\theta)$ is Borel measurable, $R < \rho(\theta) < 1$ and

$$\iint\limits_{R<|z|<1} \frac{q(z)}{|z-\rho(\theta)e^{i\theta}|}\, dx\, dy \geq c_0 > 0, \tag{6.12}$$

where $q = |\text{grad } f|$ and c_0 is independent of R and of $\theta \in G_R$. There is a distribution μ_R of unit mass on G_R with a uniformly bounded α-potential. We integrate (6.12) with respect to μ_R and use Schwarz's inequality. We find

$$c_0^2 \leq \iint\limits_{R<|z|<1} |q|^2 (1-|z|)^\alpha dx\, dy$$

$$\times \iint\limits_{G_R\, G_R} d\mu_R(\theta_1)\, d\mu_R(\theta_2) \iint\limits_{R<|z|<1} \frac{dx\, dy}{(1-|z|)^\alpha |z-\zeta_1|\, |z-\zeta_2|}$$

if $\zeta_\nu = \rho(\theta_\nu)\, e^{i\theta_\nu}$, $\nu = 1, 2$. The first factor $= \delta_R \to 0$, $R \to 1$. In the second we use Lemma 4 and find

$$c_0^2 \leq \delta_R\, C\, I(\mu_R).$$

As $R \to 1$, this implies $I(\mu_R) \to \infty$, or $C_\alpha(G_R) \to 0$, contradicting the assumption on G_R. —Theorem 3 is thus proved.

7. Theorem 3 can be used to prove the following boundary value theorem for meromorphic functions.

THEOREM 4. *Let $f(z)$ be meromorphic in $|z| < 1$ and assume*

$$\iint\limits_{|z|<1} \frac{|f'(z)|^2}{(1+|f(z)|^2)^2} (1-|z|)^\alpha\, dx\, dy < \infty, \quad 0 \leq \alpha < 1. \tag{7.1}$$

Then

$$\lim f(z), \qquad z \to e^{i\theta} \text{ non-tangentially,}$$

exists, except when θ *belongs to a set of* α*-capacity zero.*

Remark. The existence of non-tangential limits cannot be asserted for general functions. —There exists a function $\phi_\varepsilon(z)$ of class BL such that

(a) $\phi_\varepsilon(0) = 1$;

(b) $\phi_\varepsilon \equiv 0$, $|z| > \varepsilon$;

(c) $\iint |\text{grad } \phi_\varepsilon|^2 dx\, dy < \varepsilon$.

Choosing $z_{\nu n}$ on $|z| = 1 - 2^{-n}$ sufficiently dense and $\varepsilon_{\nu n}$ small enough

$$f(z) = \sum_{\nu, n} \phi_{\varepsilon_{\nu n}}(z - z_{\nu n})$$

is an example of the desired kind.

Proof. The proof consists of two steps.

1) We first observe that

$$\lim_{r \to 1} f(re^{i\theta}) \tag{7.2}$$

exists outside a set of α-capacity zero. This follows immediately from Theorem 3 since $(1 + |f|^2)^{-1}$ satisfies the assumptions of this theorem. Hence

$$\lim |f(re^{i\theta})| \text{ exists (possibly} = \infty) \text{ p.p. } (\alpha).$$

Replacing f by $f - c$, we see that also

$$\lim |f(re^{i\theta}) - c| \text{ exists p.p. } (\alpha).$$

Since the circles $|w| = a$ and $|w - c| = b$ intersect at most in two points and the clusterset of $f(re^{i\theta})$ must be one point or a continuum,

(7.2) exists if both the above limits exist, that is except in a set of α-capacity zero.

2) To prove the existence of *non-tangential* limits we consider the circles

$$C_n: \quad |z| = 1 - 2^{-n} = r_n .$$

We choose a (large) integer a, which will determine the size of the angle, and subdivide C_n into (overlapping) sectors

$$\omega_{\nu n}: \quad \nu \cdot 2\pi \cdot 2^{-n} < \theta < \nu \cdot 2\pi \cdot 2^{-n} + 2\pi a\, 2^{-n}, \quad \nu = 1, \ldots, 2^n,$$

and consider the domains

$$A_{\nu n}: \begin{cases} \theta \in \omega_{\nu n} \\ r_{n-1} < r < r_{n+2} . \end{cases}$$

If we define

$$(7.3) \qquad a_n = \sum_{\nu=1}^{2^n} \iint_{A_{\nu n}} \frac{|f'(z)|^2}{(1+|f(z)|^2)^2}\, dx\, dy \cdot 2^{-n\alpha},$$

(7.1) implies

$$\sum_1^\infty a_n = M(a) < \infty .$$

The regions $A_{\nu n}$ are divided into two groups. Assume first that

$$\iint_{A_{\nu n}} \frac{|f'(z)|^2}{(1+|f(z)|^2)^2}\, dx\, dy > 1 .$$

This holds for $m = m_n$ indices ν and by (7.3) $m < 2^{n\alpha} a_n$. We call the corresponding ω-intervals γ_{in}, $1 \le i \le m$, and denote their lengths $|\gamma_{in}|$. Then

$$\sum_{n=1}^{\infty}\sum_{\nu=1}^{m_n}|\gamma_{in}|^{\alpha} \leq (2\pi a)^{\alpha}\Sigma\, m_n 2^{-n\alpha} < (2\pi a)^{\alpha}\sum_{1}^{\infty} a_n < \infty .$$

If E_a denotes the set of θ's, included in infinitely many intervals γ_{in} the above inequality implies

$$\Lambda_{r^{\alpha}}(E_a) = 0$$

which then yields

$$C_{\alpha}(E_a) = 0 .$$

If we let a assume the values $1, 2, \ldots,$ and set $E = \bigcup_{a=1}^{\infty} E_a$, also $C_{\alpha}(E) = 0$.

We now choose a point $\theta_0 \notin E$ where (7.2) exists and assume e.g. that the limit $= 0$. We fix an integer a. For $n \geq n_0$ this point θ_0 belongs to intervals $\omega_{\nu n}$ of the second group. This means that there are regions A_n

$$A_n : \begin{cases} |\theta - \theta_0| < 2a\, 2^{-n} \\ r_{n-1} < r < r_{n+2} \end{cases}$$

such that

$$\iint_{A_n} \frac{|f'(z)|^2}{(1+|f(z)|^2)^2}\, dx\, dy \leq 1 .$$

Since the Riemann half sphere has area $\frac{\pi}{2} > 1,5$, $f(z)$ omits for $z \in A_n$ a set S_n in $|w| < 1$ of area $> 0,5$. Let us now consider the class of measurable functions on S_n satisfying $|q(w)| \leq 1$ and form the holomorphic function in A_n

$$F_q(z) = \iint_{S_n} \frac{q(w)\, du\, dv}{f(z) - w}, \quad w = u + iv .$$

Uniformly in q *and* n, $F_q(z)$ has the following properties:

(7.4) $$|F_q(z)| \leq \text{Const.}, \quad z \in A_n$$

and

(7.5) $$\left| F_q(re^{i\theta_0}) - \iint_{S_n} \frac{q(w)\, du\, dv}{-w} \right| \leq \mathcal{E}_n, \quad r_{n-1} < r < r_{n+2},$$

where $\mathcal{E}_n \to 0$ since $\lim f(re^{i\theta_0}) = 0$. By the principle of the harmonic majorant (7.4) and (7.5) imply

(7.6) $$\left| F_q(z) - \iint \frac{q(w)\, du\, dv}{-w} \right| \leq \mathcal{E}'_n, \quad |\theta - \theta_0| \leq a \cdot 2^{-n},$$

$$r_n \leq r \leq r_{n+1},$$

where $\mathcal{E}'_n \to 0$, $n \to \infty$. Taking the upper bound in (7.6) for all q, $|q| \leq 1$, we get

$$\mathcal{E}'_n \geq \sup_{|q| \leq 1} \left| \iint_{S_n} \left(\frac{1}{f(z) - w} + \frac{1}{w}\right) q(w)\, du\, dv \right| =$$

$$= \iint_{S_n} \frac{|f(z)|}{|f(z) - w|\,|w|}\, du\, dv$$

which implies

$$\lim |f(z)| = 0, \quad \begin{cases} |\theta - \theta_0| \leq a \cdot 2^{-n} \\ 2^{-n} \leq 1 - r \leq 2^{-n-1}. \end{cases}$$

Since a is arbitrarily large, the theorem is proved.

We finally note that in case $a = 0$ the proof is simplified since the first case does not occur.

8. We shall, in passing, give a geometric application of Theorem 3.

THEOREM 5. *Let $f(z)$ be a differentiable mapping of $|z| < 1$ onto a surface Y over a sphere S. Let E be a totally disconnected subset of S and assume that the part of Y situated over every compact subset of $S - E$ has finite area. Then $\lim_{r \to 1} f(re^{i\theta})$ exists except on a set of logarithmic capacity zero.*

Proof. Consider an arbitrary open spherical cap K, $\overline{K} \subset S - E$. Let $\phi(s)$ be defined on S such that $\phi \equiv 0$ outside K, $\phi > 0$ on K and let ϕ have continuous derivatives. The function $\phi(f(z))$ has a finite Dirichlet integral over $|z| < 1$ by our assumptions on f. By Theorem 3, $\phi(f(z))$ then has radial limits except on a set of logarithmic capacity zero (= p.p.). The cluster set of $f(re^{i\theta})$ along a radius for which $\phi(f)$ has limits, must be a subset of $\{s | \phi(s) = \lambda\}$ for some λ. If $\lambda > 0$ we study another function ϕ_1 so that $\phi = \lambda > 0$ and $\phi_1 = \lambda_1 > 0$ intersect in isolated points. We thus find that p.p. either

$$\lim f(re^{i\theta}) \text{ exists}$$

or

$$\text{the cluster set of } f(re^{i\theta}) \text{ is outside } K.$$

We now cover $S - E$ by caps $\{K_\nu\}_1^\infty$ and find that the above alternative holds with K replaced by $S - E$. The second alternative now means that the cluster set is a subset of E. Since E is totally disconnected the cluster set reduces to one point, i.e.,

$$\lim f(re^{i\theta}) \text{ exists.}$$

9. It is possible to generalize Fatou's theorem in another direction. Theorem 2 gives better information about convergence of Fourier series than the classical Kolmogorov-Seliverstov-Plessner on a.e. convergence

if $\lambda_n = (\log n)^{1+\delta}$, $\delta > 0$. For $\delta = 0$, however, $\overline{K}(r)$ is not integrable and Theorem 2 is empty. We can, however, consider existence of radial limits and here it is actually possible to get a complete description of the exceptional sets even in the interval $-1 < \delta < 0$. The suitable tool turns out to be the maximal functions.

Let $f(x)$ be periodic with period 2π and assume $f(x) \in L^p(-\pi, \pi)$, some $p \geq 1$. The maximal function $f^*(x)$ associated with $f(x)$ was introduced by Hardy and Littlewood through the definition

$$f^*(x) = \sup_t \frac{1}{t} \int_x^{x+t} f(u)du. \tag{9.1}$$

The inequalities

$$\int_{-\pi}^{\pi} |f^*(x)|^p dx \leq A_p \int_{-\pi}^{\pi} |f(x)|^p dx, \quad p > 1 \tag{9.2}$$

and

$$m\{x \mid f^*(x) \geq \lambda\} \leq \frac{A}{\lambda} \int_{-\pi}^{\pi} |f(x)|\, dx \tag{9.3}$$

are basic in the theory of differentiation. They can alternatively be given as theorems on harmonic functions. Assume $f > 0$ and let $u(z)$ be harmonic in $|z| < 1$ with boundary values $f(\theta)$. Then clearly

$$\text{Const. } f^*(\theta) \leq \sup_r u(re^{i\theta}) \leq \text{Const. } f^*(\theta). \tag{9.4}$$

The inequality (9.2) follows if we can characterize those non-negative measures μ for which

$$\iint_{|z|<1} u(z)^p d\mu(z) \leq A_p \int_{-\pi}^{\pi} f(x)^p dx. \tag{9.5}$$

It is sufficient to consider $p = 2$ and the complete solution is as follows: a necessary and sufficient condition on μ, is $\mu(S) \leq \text{Const.}\, s$ for every set S: $1-s < |z| < 1$, $|\arg(z) - \alpha| < s$.

The corresponding linear problem, i.e. to describe those μ for which

$$\iint u(z)\, d\mu(z) \tag{9.6}$$

is bounded for $f \in L^p$ is clearly much simpler and the solution is that

$$\phi(\theta) = \iint \frac{1-|z|^2}{|e^{i\theta} - z|^2}\, d\mu(z) \tag{9.7}$$

belongs to L^q. A solution of this problem is in principle sufficient in order to obtain results on existence of boundary values.

We shall now consider the corresponding linear problem for the class of functions $f(x)$,

$$f(x) \sim \sum_{-\infty}^{\infty} C_n e^{inx},$$

such that

$$\|f\|_K^2 = \Sigma\, |C_n|^2 \lambda_{|n|} < \infty.$$

Here $\{\lambda_n\}$ is a positive sequence such that

$$K(x) \sim \sum_0^{\infty} \frac{\cos nx}{\lambda_n}$$

is a convex function $\in L^1$. The following theorem is quite easy to prove.

THEOREM 6. *If* $\lambda_n = (n+1)^{1-\alpha}$, $0 \leq \alpha < 1$, (9.6) *is bounded if and only if*

$$E_\alpha(\mu) = \iint \frac{d\mu(a)\, d\mu(b)}{|1 - a\bar{b}|^\alpha} < \infty, \quad 0 < \alpha < 1,$$

$$E_0(\mu) = \iint \log \left| \frac{1}{1-\bar{a}\, b} \right| d\mu(a)\, d\mu(b) < \infty\,, \quad \alpha = 0\,.$$

The bound of (9.6) is $\leq$ Const. $\sqrt{E_\alpha}$.

If we specialize $d\mu$ to have the form $d\sigma(\theta)$ placed at a point on the radius from 0 to $e^{i\theta}$ we find using (9.4) and observing that $E_\alpha(\mu)$ essentially increases if we push the masses out to $|z| = 1$

$$\left(\int f^*(x)\, d\sigma(x) \right)^2 \leq A_\alpha \, \|f\|_K^2 \, I_\alpha(\sigma) \tag{9.8}$$

where I_α is the energy of σ with respect to the kernel $|x|^{-\alpha}$, resp. $\log \frac{1}{|x|}$. This inequality implies the existence of derivatives and boundary values except on sets of capacity zero, as also follows from 2.

We give the proof of Theorem 6 only in the case $\alpha = 0$ which is particularly simple. Consider first the case when μ has its support strictly inside $|z| < 1$. Consider the harmonic function

$$u_0(z) = \iint \log |1 - z\bar{\zeta}|\; d\mu(\zeta)$$

and let (u, v) denote scalar product in the space of harmonic functions with finite Dirichlet integral and with $u(0) = 0$. Then by Poisson's formula

$$(u, u_0) = \int_{|z|=1} u \frac{\partial u_0}{\partial n}\, ds = 2\pi\, u(z)\, d\mu(z)\,.$$

Hence

$$2\pi \left| \iint u \, d\mu \right| \leq \|u_0\| \cdot \|u\|$$

with equality if $u = u_0$, and the linear functional (9.6) has norm $(2\pi)^{-1/2} \sqrt{E_0(\mu)}$. The case of a general μ follows immediately. If $u(0) = \frac{1}{2\pi} \int f dx \neq 0$, we simply consider $u - u(0)$.

10.   It is clearly possible to use the same method for general kernels $K(x)$ and corresponding weights λ_n. However, the formulas become so involved that they cannot be used to deduce inequalities of the form (9.8). The case

$$\lambda_n = (\log (n + 2))^\alpha, \qquad 0 < a < \infty,$$

is particularly interesting. The kernel K_α that is associated with this sequence is

$$K_\alpha(x) \sim \frac{1}{|x| (\log \frac{1}{|x|})^{1+a}}, \qquad x \to 0.$$

The following theorem holds.

THEOREM 7.   *There is a constant* B_α *such that*

$$C_{K_\alpha} [\{x \mid f^*(x) \geq \lambda\}] \leq \frac{B_\alpha}{\lambda^2} \|f\|_{K_\alpha}^2, \qquad 0 < a < \infty.$$

Remark.   By standard methods this implies that the primitive function of f has a derivative and that the corresponding harmonic function has boundary values except on sets of K_α-capacity zero.

In the proof we use the equivalent norm

$$(10.1) \qquad \int\!\!\int_{-\pi}^{\pi} \frac{|f(x) - f(y)|^2}{\phi(x - y)} \, dx \, dy, \qquad \phi(t) = |t| (\log \frac{8}{|t|})^{1-a}$$

and the following potential theoretic lemma:

LEMMA 6.   *If* σ *is an interval of length* d *on* $(-\pi, \pi)$, *denote by* $T\sigma$ *an interval of length* $3d$ *and the same midpoint as* σ. *We assume that* $\{\sigma_\nu\}$ *are disjoint and denote by* $E = \cup \sigma_\nu$ *and* $E' = \cup T\sigma$. *There is a constant* Q *only depending on* K *such that*

$$C_K(E') \leq Q \, C_K(E)$$

provided $K(x) = O(K(2x)),\ x \to 0$.

In an outline, the proof of Theorem 2 proceeds as follows. Let $\sigma_{\nu n}$ denote the 2^n disjoint intervals of length $2\pi \cdot 2^{-n}$ on $(-\pi, \pi)$. Let λ be given and denote by $M_\alpha(f)$ the mean value of f over the interval α. We choose intervals $\sigma_1, \sigma_2, \ldots,$ such that

$$M_{\sigma_\nu}(f) \geq \lambda \tag{10.2}$$

by first choosing those $\sigma_{\nu 1}$ that satisfy (10.2), then $\sigma_{\mu 2}$ disjoint from those chosen before etc. It follows easily from the lemma that it is sufficient to prove $C\{\cup \sigma_\nu\} \leq \text{Const.}\ \|f\|^2 \cdot \lambda^{-2}$.

Let τ_ν be intervals such that $T\tau_\nu = \sigma_\nu$. We want to construct $f_1(x)$ such that $\|f_1\| \leq \text{Const.}\ \|f\|$ and $f_1(x) \equiv M_{\sigma_\nu}(f),\ x \in \tau_\nu$. We first modify f on each σ_ν according to the following rule where we have normalized σ_ν to $(-1, 1)$:

$$f_2(x) = \begin{cases} f(2x), & -\frac{1}{2} < x < \frac{1}{2} \\ f(-x - \frac{3}{2}), & -\frac{3}{4} < x < -\frac{1}{2} \\ f(x), & -1 < x < -\frac{3}{4} \\ \text{analogously on } (\frac{1}{2}, 1). & \end{cases}$$

Outside $\cup \sigma_\nu$ we define $f_2(x) = f(x)$. From (10.1) it follows that $\|f_2\|_K \leq \text{Const.}\ \|f\|_K$.

Let 4δ be the length of the shortest of the intervals σ_ν. We have the following picture:

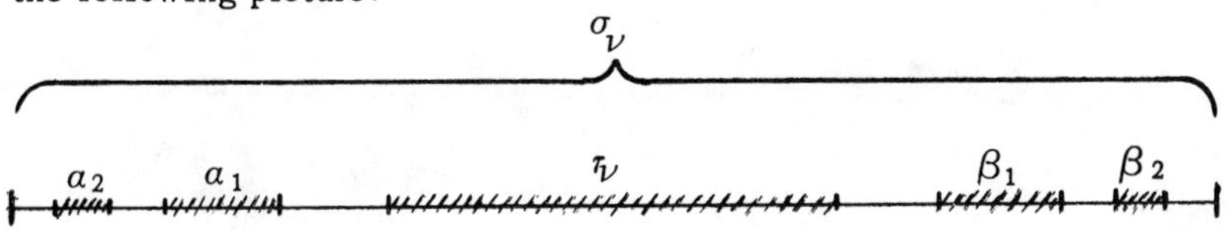

where we construct α_i and β_i until their length $< \delta$. We define

$$f_1(x) = \begin{cases} M_{\tau_\nu}(f_2) = M_{\sigma_\nu}(f), & x \in \tau_\nu; \\ M_{\alpha_i}(f_2), & x \in \alpha_i; \\ M_{\beta_i}(f_2), & x \in \beta_i; \\ \text{linear between the intervals.} \end{cases}$$

We do the same construction on each σ_ν and each complementary interval. A computation in (10.1) shows that $\|f_1\| < \text{Const.}\ \|f_2\|$.

To complete the proof, let μ be a distribution of unit mass on $E'' = \cup \tau_\nu$. Then

$$\lambda \leq \int_{E''} f_2(x)\, d\mu(x) \leq \|f_2\|_K \cdot I_K(\mu)^{\frac{1}{2}} \leq \text{Const.}\ \|f\|_K \cdot I_K(\mu)^{\frac{1}{2}}.$$

The lemma now yields Theorem 7.

§ VI. *EXISTENCE OF CERTAIN HOLOMORPHIC FUNCTIONS*

1. Let D be a connected domain in the complex z-plane, containing the point at ∞. We consider the space $H^q(D)$, $1 \leq q \leq \infty$, of holomorphic functions f in D satisfying

$$(1.1) \qquad \|f\|_q = \left\{ \iint_D |f(z)|^q dx\, dy \right\}^{\frac{1}{q}} < \infty, \quad \|f\|_\infty = \sup_{z \in D} |f(z)|.$$

The compact complement of D is denoted E. Our problem in this section is to find metrical conditions on E which guarantee that $H^q(D)$ contains non-trivial functions f. We observe that the case $q < 2$ is trivial since $(z - z_0)^{-1} - (z - z_1)^{-1}$, $z_i \in E$, always belongs to $H^q(D)$, $q < 2$. The following theorem will be proved.

THEOREM 1.

(a) $H^2(D)$ *contains non-trivial functions if and only if* $C_0(E) > 0$.

(b) $H^q(D)$ *contains non-trivial functions if* $C_{2-p}(E) > 0$, where $p^{-1} + q^{-1} = 1$, $2 < q \leq \infty$.

(c) $H^q(D)$ *contains only* $f \equiv 0$ *if* $\Lambda_{2-p}(E) < \infty$, $q < \infty$, $\Lambda_1(E) = 0$, $q = \infty$.

Proof.

(a) 1) We first assume $C_0(E) > 0$. E has non-intersecting closed subsets E_ν such that $C_0(E_\nu) > 0$, $\nu = 1, 2$. Let μ_ν be distributions of unit mass on E_ν with bounded logarithmic potentials and form $\mu = \mu_1 - \mu_2$ and

$$f(z) = \int_E \frac{d\mu(\zeta)}{\zeta - z} = \frac{a}{z^2} + \ldots, \quad |z| > R.$$

Clearly

$$\iint_{|z|>R} |f(z)|^2 dx\, dy < \infty, \quad R \text{ large enough,}$$

while

$$\iint_{|z| \leq R} |f(z)|^2 dx\, dy \leq \int_E \int_E |d\mu(\zeta)|\, |d\mu(\zeta')| \iint_{|z| \leq R} \frac{dx\, dy}{|z-\zeta'|\, |z-\zeta|} \leq$$

$$\leq \text{Const.} \int_E \int_E \log \left|\frac{3R}{\zeta - \zeta'}\right| |d\mu(\zeta)|\, |d\mu(\zeta')| < \infty.$$

To see that $f(z) \not\equiv 0$ choose a system γ of analytic curves enclosing E_1. Then

$$\frac{1}{2\pi i} \int_\gamma f(z)\, dz = -\mu_1(E_1) = -1.$$

2) Assume $C_0(E) = 0$ and choose $D_1 \subset D_2 \subset \ldots \to D$, where D_ν is bounded by a finite number of analytic curves. Let g_ν be the Green's function of D_ν with pole at ∞ and let h_ν be its conjugate function. It is well known (see also next chapter) that $C_0(E) = 0$ is equivalent to $g_\nu(z) \to \infty$, $z \in D$, uniformly on inside domains. Hence if we introduce the coordinates $\zeta = \xi + i\eta = g_\nu + ih_\nu$ in D_ν and consider the domain $\Omega_\nu = \{z \mid 0 < g_\nu < 1\}$ we have for $f \in H^2(D)$,

$$\iint_{\Omega_\nu} |f(z)|^2 dx\, dy = \int_0^1 d\xi \int_0^{2\pi} \frac{|f|^2 d\eta}{\left(\frac{\partial g_\nu}{\partial n}\right)^2} = \varepsilon_\nu \to 0, \quad \nu \to \infty.$$

On the other hand, the following estimate holds:

$$\int_0^1 d\xi\Big(\int_{g_\nu=\xi} |f(z)|\,|dz|\Big)^2 \le \int_0^1 d\xi\left\{\int_{g_\nu=\xi}\frac{\partial g_\nu}{\partial n}|dz|\cdot\int_{g_\nu=\xi}\frac{|f|^2}{\frac{\partial g_\nu}{\partial n}}|dz|\right\} =$$

$$= 2\pi\int_0^1\int_0^{2\pi}\frac{|f|^2}{\left(\frac{\partial g_\nu}{\partial n}\right)^2}\,d\eta\,d\xi = 2\pi\,\varepsilon_\nu .$$

Hence ξ_ν, $0 < \xi_\nu < 1$, exists so that

$$\int_{g_\nu=\xi_\nu} |f(z)|\,|dz| \le \sqrt{2\pi\,\varepsilon_\nu}\,.$$

The curves $g_\nu = \xi_\nu$ approach E uniformly as $\nu \to \infty$. Let z_0 be fixed in D. Then for $\nu > \nu_0$ since $f(\infty) = 0$

$$|f(z_0)| = \left|\frac{1}{2\pi i}\int_{g_\nu=\xi_\nu}\frac{f(z)dz}{z-z_0}\right| \le \text{Const.}\cdot\sqrt{2\pi\,\varepsilon_\nu} \to 0\,.$$

This implies $f(z) \equiv 0$ and concludes the proof of (a).

(b) Assume $C_{2-p}(E) > 0$ and choose $\mu > 0$ with support on E so that

$$\int_E \frac{d\mu(\zeta)}{|z-\zeta|^{2-p}} \le M\,. \tag{1.2}$$

We shall prove that

$$f(z) = \int_E \frac{d\mu(\zeta)}{z-\zeta} = \frac{\mu(E)}{z} + \cdots \qquad (\neq 0)$$

belongs to $H^q(D)$. We assume $q < \infty$, $q = \infty$ being trivial.

Let $t(z)$ be a step function (= constant on each rectangle in a net) such that $t(z) > 0$ and

$$\iint_D t(z)^p dx\,dy = 1\,. \tag{1.3}$$

$s = \sigma + it$ denotes a complex variable, $0 \le \sigma \le 1$, and we form

$$\phi(s) = \iint\limits_D \int\limits_E \frac{t(z)^{p-\frac{1}{2}ps}}{|z-\zeta|^{\frac{1}{2}ps+2-p}}\, d\mu(\zeta)dx\, dy\,.$$

By (1.2) and (1.3) we have $|\phi(it)| \le M$. For $\sigma = 1$ we find by Schwarz's inequality

$$|\phi(1+it)| \le \Big(\iint\limits_D t(z)^p dx\, dy\Big)^{\frac{1}{2}} \int\limits_E\int\limits_E d\mu(\zeta)d\mu(\zeta') \cdot \iint\limits_{-\infty}^{\infty} \frac{dx\, dy}{|z-\zeta|^k |z-\zeta'|^k}$$

where $k = 2 - \frac{1}{2}p$. The last integral is $=$ Const. $|\zeta-\zeta'|^{p-2}$ and so

$$|\phi(1+it)| \le \text{Const.}\, M^{\frac{1}{2}} < \text{Const.}\, M\,.$$

By the maximum principle we have $\phi(2-\frac{2}{p}) <$ Const. M which implies

$$\iint\limits_D t(z)\,|f(z)|\, dx\, dy < \text{Const.}\, M\,.$$

Here $t(z)$ satisfying (1.3) is arbitrary so by the converse of Hölder's inequality

$$\iint |f(z)|^q dx\, dy < \infty\,.$$

(c) Assume first $q < \infty$ and that $\Lambda_{2-p}(E) = \Lambda < \infty$. Let $\{C_\nu\}_1^\infty$ be a covering of E with circles

$$C_\nu:\quad |z - z_\nu| < r_\nu$$

such that

(1.4) $$\Sigma\, r_\nu^{2-p} \le 2\,\Lambda\,, \qquad r_\nu \le \rho\,.$$

t is a parameter $1 \le t \le 2$ and we consider the expanded circles

$$C_{\nu t}:\quad |z - z_\nu| < r_\nu t\,, \quad 1 \le t \le 2\,.$$

The boundary of the unbounded component of the complement of $\bigcup_{\nu=1}^{n} C_{\nu t}$ is denoted Γ_t and consists of certain arcs $b_\nu(t)$ of the circumferences of $C_{\nu t}$. As t varies, $1 \leq t \leq 2$, $b_\nu(t)$ sweeps out a set Ω_ν. $\partial\Omega_\nu$ is piecewise a circle and we observe that Ω_ν and Ω_μ have no interior points in common. To see this, let z_0 be an interior point of Ω_ν and assume that $z_0 \in b_\nu(\tau)$. For $t > \tau$, z_0 is an interior point of $C_{\nu t}$ and hence $z_0 \notin \Gamma_t$, $t > \tau$, and so $z_0 \notin b_\mu(t)$.

Let $f(z) \in H^q(D)$. Then

$$f'(\infty) = \frac{1}{2\pi i} \int_{\Gamma_t} f(z)dz = \frac{1}{2\pi i} \int_1^2 dt \int_{\Gamma_t} f(z)dz =$$

$$= \frac{1}{2\pi i} \sum_{\nu=1}^{n} \int_0^1 dt \int_{b_\nu(t)} f(z)dz .$$

Taking absolute values, we find

$$|f'(\infty)| \leq \frac{1}{2\pi} \sum_{\nu=1}^{n} \int_1^2 dt \int_{b_\nu(t)} |f(z)|\,|dz| \leq \frac{1}{2\pi} \sum_{\nu=1}^{n} r_\nu^{-1} \iint_{\Omega_\nu} |f(z)|\,dx\,dy \leq$$

$$\leq \frac{1}{2\pi} \sum_{\nu=1}^{n} r_\nu^{-1} \Big(\iint_{\Omega_\nu} |f(z)|^q dx\,dy\Big)^{\frac{1}{q}} \Big(\iint_{\Omega_\nu} dx\,dy\Big)^{\frac{1}{p}} \leq \tag{1.5}$$

$$\leq \frac{1}{2\pi} \sum_{\nu=1}^{n} r_\nu^{-1} a_\nu (3\pi\, r_\nu^2)^{\frac{1}{p}} \leq \frac{(3\pi)^{\frac{1}{p}}}{2\pi} (\Sigma\, a_\nu^q)^{\frac{1}{q}} \cdot (\Sigma\, r_\nu^{2-p})^{\frac{1}{p}},$$

where $a_\nu{}^q$ denotes the double integral of $|f|^q$ over Ω_ν. Since Ω_ν are (essentially) disjoint, and since $\{\Omega_\nu\}$ cover a small neighbourhood D_ρ of E, we have

$$\Sigma\, a_\nu^q = \iint_{D_\rho} |f(z)|^q dx\,dy = \varepsilon(\rho) \to 0, \quad \rho \to 0 .$$

(1.5) and (1.4) then yield

$$|f'(\infty)| \leq \text{Const.}\ (2\Lambda)^{\frac{1}{p}} \epsilon(\rho)^{\frac{1}{q}} \to 0, \quad \rho \to 0.$$

Hence $f'(\infty) = 0$. Considering in the same way $zf(z)$, $z^2 f(z), \ldots$ we find $f \equiv 0$.

(c) $q = \infty$. We can enclose E by circles C_ν so that

$$\Sigma r_\nu < \varepsilon.$$

Then

$$|f'(\infty)| \leq \frac{1}{2\pi} \Sigma \int_{C_\nu} |f(z)|\,|dz| \leq \|f\|_\infty \cdot \varepsilon$$

and as above we find $f \equiv 0$.

2. The space $H^\infty(D)$ is the most important case in the above Theorem 1. The existence of a non-trivial function is related to the extremal problem of maximizing the coefficient a in the development

$$f(z) = \frac{a}{z} + \ldots$$

where $\|f\|_\infty \leq 1$. The extremal function $f_\infty(z) = \frac{a}{z} + \ldots$ is conveniently discussed as limit of the corresponding extremal function for the space $H^q(D)$, as $q \to \infty$. We shall briefly outline this procedure and some of its consequences.

Consider the extremal problem

$$\text{(2.1)} \qquad \inf \|f\|_q = m_q, \quad f(z) = \frac{1}{z} + \ldots,$$

where $f \in H^q(D)$ and ∂D is assumed to consist of n analytic curves. The space H^q being uniformly convex, there exists an extremal function $f = F_q$. Let $g(z) = \beta z^{-2} + \ldots$ be analytic in D and consider $F_q + tg$. Varying the parameter t we get

(2.2) $$\iint_D |F_q|^{q-2}\,\overline{F}_q\, g\, dx\, dy = 0 .$$

We set

$$f_q(z) = m_q^{-1}\, F_q(z) = \frac{m_q^{-1}}{z} + \ldots$$

and

$$\phi_q(z) = |F_q(z)|^{q-2}\,\overline{F}_q(z)\, m_q^{1-q} \; |f_q|^{q-2}\,\overline{f}_q .$$

If $D_0 \subset D$ is a bounded domain, (2.1) and Hölder's inequality yield

(2.3) $$\|\phi_q\|_p = 1 , \quad \overline{\lim_{q\to\infty}} \iint_{D_0} |\phi_q(z)|\, dx\, dy \le 1 ,$$

while

(2.4) $$\iint_D \phi_q(z)\, f_q(z)\, dx\, dy = 1 .$$

If $h = a \cdot z^{-1} + \ldots$ belongs to H^q it is easy to see, by approximations by functions analytic in $\overline{D}$, that (2.2) holds for $g = h - a m_q f_q$ which by (2.4) implies

(2.5) $$\iint_D \phi_q(z)\, h(z)\, dx\, dy = a\, m_q .$$

We now let $q = q_i \to \infty$ so that $f_{q_i} \to f_\infty$ and $\phi_{q_i} dx\, dy$ tends weakly to a set function $d\mu$. f_∞ is then an extremal function for $H^\infty(D)$, $|f_\infty| \le 1$, and the definition of ϕ_q and the maximum principle show that $S_\mu \subset \partial D$. (2.3) implies

(2.6) $$\int_{\partial D} |d\mu| \le 1 .$$

(2.5) implies if $h(z)$ is continuous in $\overline{D}$

(2.7) $$\int_{\partial D} h(z)\, d\mu(z) = a\, a^{-1} ,$$

since $\lim_{q\to\infty} m_q = a^{-1}$. Choosing $h(z) = (z-\zeta)^{-1}$, $\zeta \notin \bar{D}$, (2.7) yields

$$\int_{\partial D} \frac{d\mu(z)}{z-\zeta} \equiv a^{-1}, \quad \zeta \notin \bar{D}.$$

By the famous Riesz theorem, $d\mu(z) - \phi(z)\,dz$ where $\phi(z)$ are boundary values of a function analytic in D and

$$\int_{\Gamma_\lambda} |\phi(z)|\,|dz| \leq 1, \tag{2.8}$$

where Γ_λ are the level curves $g = \lambda$ of the Green's function of D, and finally $a^{-1} = 2\pi i\,\phi(\infty)$. Since ϕ tends strongly to its boundary values, (2.7) holds for any h, $|h| \leq 1$. We also observe that if ψ is any function satisfying (2.8) we have

$$|a\psi(\infty)| = \left|\frac{1}{2\pi i}\int_{\Gamma_\lambda} \psi\, f_\infty\, dz\right| \leq \frac{1}{2\pi}. \tag{2.9}$$

Hence $(2\pi a)^{-1}$ *is* sup $|\psi(\infty)|$ *for all* ψ, *satisfying (2.8).*

Since we have equality in (2.9) for $\psi = \phi$ we can conclude (a) that $|f_\infty(z)| = 1$ a.e. on ∂D and (b) that $\phi f_\infty dz \geq 0$ on ∂D. Since the boundary ∂D is analytic we deduce that (c) ϕf_∞ is analytic on ∂D and (d) that ϕf_∞ has exactly n zeros in $\bar{D}$. In order to prove that $|f_\infty(z)| \equiv 1$ on ∂D we have to prove that

$$\int_{e_\lambda} \log|f_\infty(z)|\,|dz| \to 0, \text{ if } \int_{e_\lambda} |dz| \to 0,\ \lambda \to 0,\ e_\lambda \subset \Gamma_\lambda.$$

We have

$$\begin{aligned}\int_{e_\lambda} \log|f_\infty(z)|\,|dz| &= \int_{e_\lambda} \log|f_\infty\phi|\,|dz| - \int_{e_\lambda} \log|\phi|\,|dz| \geq \\ &\geq o(1) - \int_{e_\lambda} \overset{+}{\log}|\phi|\,|dz| \geq \\ &\geq o(1) - \int_{e_\lambda} |\phi|^{\frac{1}{2}}\,|dz| \geq\end{aligned}$$

$$\geq o(1) - \left(\int_{e_\lambda} |dz| \right)^{\frac{1}{2}} \int_{e_\lambda} |\phi| \, |dz| = o(1).$$

Hence (e) $|f_\infty(z)| \equiv 1$ *on* ∂D *and* f_∞ *is analytic on* ∂D, which then by (c) yields (f) that ϕ *is analytic on* ∂D. (e) implies that $f_\infty(z)$ has at least n zeros and so by (d) that (g) $f_\infty(z)$ *has exactly* n *zeros and* $\phi \neq 0$ *in* $\bar{D}$.

Now let D be an arbitrary domain for which $H^\infty(D)$ contains non-trivial functions. Then $f \in H^\infty$ exists with $a \neq 0$. Let D_n be approximations to D, $D_1 \subset D_2 \subset \ldots \to D$, and let $f_n(z) = a_n z^{-1} + \ldots$ be an extremal function of D_n. Clearly $a_n \geq a > 0$ and $\lim a_n = a$. We apply (2.7) to the function

$$\frac{f_n(z) - f_n(\zeta)}{z - \zeta} = -\frac{f_n(\zeta)}{z} + \ldots, \quad \zeta \in D_n,$$

and find

$$f_n(\zeta) = \int_{\partial D_n} \frac{f_n(z)\, \phi_n(z)\, dz}{z - \zeta} \cdot \left\{ \int_{\partial D_n} \frac{\phi_n(z)}{z - \zeta} dz - a_n^{-1} \right\}^{-1}.$$

We choose $n_i \to \infty$ so that $f_{n_i} \phi_{n_i} dz = |\phi_{n_i}| \, |dz| \to dp$, a distribution of unit mass on ∂D. $f_{n_i} \to f$, an extremal function for D and

$$f(\zeta) = \int_{\partial D} \frac{dp(z)}{z - \zeta} \{\Phi(z) - a^{-1}\}, \tag{2.10}$$

where $\Phi(\infty) = 0$. Now let $f_\infty(z) = az^{-1} + \ldots$ be an arbitrary extremal function for D and apply (2.7) for $h = f_\infty$ and the domain D_n:

$$f_\infty(z) = \int_{\partial D_n} \frac{f_\infty(z)\, \phi_n(z) dz}{z - \zeta} \cdot \left\{ \int_{\partial D_n} \frac{\phi_n(z)}{z - \zeta} dz - a_n^{-1} \right\}^{-1}.$$

Since $|f_\infty \phi_n dz| \leq |\phi_n| \, |dz|$, while on the other hand

$$\int_{\partial D_{n_i}} f_\infty \phi_{n_i} dz \to 1,$$

it follows that

$$f_\infty \phi_{n_i} dz \to dp .$$

We thus obtain the same representation (2.10) for f_∞ as for f. Hence *the extremal* $f_\infty(z)$ *is unique even in the most general case.*

3. We shall finally discuss the existence of analytic functions with finite Dirichlet integral outside a set E situated on a smooth curve Γ and shall prove the following theorem. To avoid the complications depending on the change of sign of $\log \frac{1}{r}$, we use the classical $C_0'(E) = \exp\{C_0(E)^{-1}\}$ with obvious definition if E is large.

THEOREM 2. *Let* E *be a closed subset of a simple closed curve* Γ *with continuously varying curvature. Outside* E *there exists a non-trivial analytic function* $f(z)$ *with finite Dirichlet integral if and only if* $C_0'(\Gamma - E) < C_0'(\Gamma)$. *This inequality is thus independent of the choice of* Γ.

Proof.

1) We first assume that Γ is the unit circle ω.

a) Assume $C_0'(\omega - E) < C_0'(\omega) = 1$.

We choose finite sums of intervals $F_n \nearrow \omega - E$ and let $u_n(z)$ be the equilibrium potentials of F_n of distributions μ_n of unit masses on F_n, $u_n \equiv \gamma_n$ on F_n. We form

$$f_n(z) = \frac{1}{2\pi} \int_{-\pi}^{\pi} \frac{e^{i\phi} + z}{e^{i\phi} - z} (u_n(e^{i\phi}) - \gamma_n) d\phi = \frac{1}{2\pi} \int_{\omega - F_n} .$$

$f_n(z)$ is holomorphic outside $\omega - F_n$, $f_n(0) = -\gamma_n$ and $f_n(\infty) = \gamma_n$. We develop $f_n(z)$ in series:

$$f_n(z) = \begin{cases} -\gamma_n + 2 \sum\limits_{\nu=1}^{\infty} z^{\nu} \dfrac{1}{2\pi} \displaystyle\int_{-\pi}^{\pi} u_n(e^{i\phi}) e^{-i\nu\phi} d\phi = \sum\limits_{0}^{\infty} a_\nu z^\nu, & |z| < 1, \\ \gamma_n - 2 \sum\limits_{\nu=1}^{\infty} z^{-\nu} \dfrac{1}{2\pi} \displaystyle\int_{-\pi}^{\pi} u_n(e^{i\phi}) e^{i\nu\phi} d\phi = \sum\limits_{0}^{\infty} b_\nu z^{-\nu}, & |z| > 1. \end{cases}$$

Observing that

$$a_\nu = \frac{1}{\nu} \int_{-\pi}^{\pi} e^{-i\nu\phi} \mu_n(\phi),$$

we find

$$\iint\limits_{|z|<1} |f_n'(z)|^2 \, dx\, dy = \pi \sum_1^\infty \nu |a_\nu|^2 = \pi I(\mu_n) < M'.$$

The integral over $|z| > 1$ is estimated in the same way and hence

$$\iint |f_n'(z)|^2 \, dx\, dy < M.$$

The function $f(z) = \lim f_{n_\nu}(z)$ is thus non-trivial, since $\lim \gamma_n > 0$, and has finite Dirichlet integral.

b) We now assume that $C_0(\omega - E) = 1$ and construct F_n and μ_n as above. Since, if $d\mu_{n_\nu} \to d\mu$, $u_\mu \le 0$ and the equilibrium distribution for ω is unique we must have

$$d\mu_n \to \frac{1}{2\pi} d\theta, \quad n \to \infty,$$

whence

$$\mu_\nu^{(n)} = \int_{-\pi}^{\pi} e^{i\nu\phi} d\mu_n(\phi)$$

have the properties

$$\mu_0^{(n)} = 1$$

and

$$\sum_{\nu\neq 0} \frac{1}{|\nu|} |\mu_\nu^{(n)}|^2 \to 0, \quad n \to \infty .$$

If $f(z)$ satisfies the conditions of the theorem, then

$$u(z) = \text{Re}\,\{f(z) - \overline{f(\bar{z}^{-1})}\} = \sum_{-\infty}^{\infty} c_\nu r^{|\nu|} e^{i\nu\theta}, \quad r < 1,$$

is harmonic outside E and $\Sigma\, |\nu|\, |c_\nu|^2 < \infty$. On F_n, $u(e^{i\theta}) = 0$ whence

$$0 = \lim_{r\to 1} \int_{F_n} u(re^{i\theta})\, e^{im\theta}\, d\mu_n(\theta) = \sum_\nu c_{\nu - m}\, d\mu_\nu^{(n)} =$$

$$= c_{-m} + \sum_{\nu\neq 0} c_{\nu-m}\, \mu_\nu^{(n)} .$$

This yields

$$|c_{-m}|^2 \le \sum_{\nu\neq 0} |\nu|\, |c_{\nu-m}|^2 \cdot \sum_{\nu\neq 0} \frac{1}{|\nu|} |\mu_\nu^{(n)}|^2 \to 0, \quad n \to \infty .$$

Hence $u(z) \equiv 0$ and if $(z) \equiv \overline{f(\bar{z}^{-1})} + ic_2$. In the same way $Im\{f(z) + f(\bar{z}^{-1})\}$ is shown to be $\equiv 0$ and we find $f \equiv$ constant.

2) To be able to prove Theorem 3 for a general Γ we need a reformulation of the condition $C_0'(\Gamma - E) = C_0'(\Gamma)$. The new condition shows more clearly which properties of E that are essential. The result is formulated in the following lemma.

LEMMA 1. *Let Γ be a curve as stated in Theorem 3. Then $C_0'(\Gamma - E) = C_0'(\Gamma)$ is equivalent to the following condition. There exist mass distributions λ_n on $\Gamma - E$ so that*

(3.1) $\lambda_n \to$ *arc length* s *(weakly)*

and

(3.2) $u_{\lambda_n}(z)$ *converge uniformly to* u_s *on* Γ.

Proof. Assume first $C_0'(\Gamma - E) = C_0'(\Gamma)$. The equilibrium distribution μ of Γ is

$$d\mu = \frac{1}{2\pi}\frac{\partial g}{\partial n}\, ds\,,$$

where g is the Green's function. For Γ, $\frac{\partial g}{\partial n}$ is bounded from above and below. Now let μ_n be equilibrium distributions of subsets of $\Gamma - E$ so that $u_{\mu_n} \leq \gamma + \varepsilon_n$, $\varepsilon_n \to 0$. Since $\mu_n \to \mu$, the above formula for μ and $u_\mu \equiv \gamma$ on Γ show that

$$\int_{|z-\zeta|>\delta} \log\frac{1}{|z-\zeta|}\, d\mu_n(\zeta) > \gamma - \varepsilon\,,$$

uniformly for $z \in \Gamma$ if $\delta < \delta(\varepsilon)$ and $n \geq n(\varepsilon)$. Hence

$$\int_{|z-\zeta|\leq\delta} \log\frac{1}{|z-\zeta|}\, d\mu_n(\zeta) < 2\varepsilon\,, \quad n \geq n_1(\varepsilon). \tag{3.3}$$

Then $d\lambda_n = 2\pi\left(\frac{\partial g}{\partial n}\right)^{-1} d\mu_n$ also have the property (3.3), which proves one part of the lemma.

Assume now that (3.1) and (3.2) hold. Then $d\mu_n = \frac{1}{2\pi}\left(\frac{\partial g}{\partial n}\right) d\lambda_n$ also generate uniformly convergent potentials from which immediately follows that $C_0'(\Gamma - E) = C_0'(\Gamma)$.

Remark. It is clear from the proof that the lemma holds for all s for which $\frac{d\mu}{ds}$ is bounded from above and from below.

We now prove Theorem 2 by showing that the properties involved are invariant under a suitable mapping of $\Gamma \leftrightarrow \omega$.

We may assume that Γ has length $= 2\pi$ and map $\Gamma \leftrightarrow \omega$ so that the arc length between corresponding points is equal. The set E corresponds to $F \subset \omega$ and the lemma shows that $C_0'(\Gamma - E) < C_0'(\Gamma)$

if and only if $C_0'(\omega - F) < C_0(\omega)$.

We shall now prove that a non-trivial function f with finite Dirichlet integral exists outside F if and only if such a function exists outside E. The proof is the same in both directions. We assume $f(z)$ given outside E and let $z = \phi(\zeta)$ be the above mapping $\omega \to \Gamma$. For a function $g(\zeta)$ with finite Dirichlet integral outside a set $F_1 \subset \omega$, boundary values $g_1(e^{i\theta})$ and $g_2(e^{i\theta})$ from the inside and the outside exist. These functions belong to all Lebesgue classes $L^q(\omega)$ and can be used in the Cauchy formula. The same result holds for Γ as is shown by a conformal mapping. Hence $f(z)$ can be written (assuming $f(\infty) = 0$)

$$f(z) = \int_E \frac{F(z_1)}{z_1 - z}\, dz_1 = \int_\Gamma \frac{F(z_1)}{z_1 - z}\, dz_1 \ (F \equiv 0,\ z_1 \notin E), \tag{3.4}$$

where $F(z_1) \in L^q(|dz_1|)$, all $q < \infty$. We construct

$$g(\zeta) = \int_\omega \frac{F(\phi(\zeta_1))\, d\zeta_1}{\zeta_1 - \zeta}. \tag{3.5}$$

It is clear that $g(\zeta)$ is holomorphic outside F and that $g(\zeta) \not\equiv 0$ (since the set function $F(\phi)\, d\zeta_1 \not\equiv 0$; cf. the proof of Theorem 1 (a)). If we can prove that g has finite Dirichlet integral, Theorem 2 is proved.

We extend the mapping $z = \phi(\zeta)$ to e.g. the inside of ω by associating to a point ζ_0 on a certain normal to ω a point inside Γ on the corresponding normal to Γ at the same distance to Γ. Our assumption on Γ implies that the extended mapping is $1-1$ in a neighbourhood of ω and that the area elements have a quotient bounded from above and from below. Let $\zeta_0 = -i + i\delta$ and $z_0 = i\delta$ be corresponding points and let the normals meet ω resp. Γ in $-i$ resp. 0. We shall compare $g'(\zeta_0)$ and $f'(z_0)$.

In the formula (3.4) we replace z_1 by x_1 and dz_1 by dx_1 and obtain a function $f_1(z)$. We find

$$|f'(z_0)-f_1'(z_0)| \le \int_\Gamma \frac{|F(z_1)|\,|dy_1|}{|z_0-z_1|^2} + \int_\Gamma |F(z_1)| \left|\frac{1}{(z_0-z_1)^2} - \frac{1}{(z_0-x_1)^2}\right| dx_1$$

$$\le C \int_\Gamma \frac{|F(z_1)|\,|x_1|\,|dx_1|}{\delta^2+x_1^2} + C\int_\Gamma |F(z_1)| \frac{\delta x_1^2 + |x_1|^3}{(\delta^2+x_1^2)^2}\,|dx_1| \tag{3.6}$$

$$= O(\delta^{-a}),\ \text{all}\ a>0,$$

since $F \in L^q(|dz_1|)$. A corresponding inequality holds for $g_1(\zeta_0)$ where in (3.5) ζ_1 and $d\zeta_1$ are replaced by ξ_1 and $d\xi_1$, $\zeta_1 = \xi_1 + i\eta_1$. We finally have the inequality

$$|f_1'(z_0)-g_1'(\zeta_0)| \le C \cdot \int |F(z_1)|\,|(x_1-i\delta)^{-2}-(x_1+O(x_1^2)-i\delta)^{-2}|\,|dx_1|$$

$$= O(\delta^{-a}),\ a>0. \tag{3.7}$$

(3.6) and (3.7) and the remarks made above now show that $g(\zeta)$ has finite Dirichlet integral. Theorem 2 is thus proved.

§ VII. *REMOVABLE SINGULARITIES FOR HARMONIC FUNCTIONS*

1. The problem to be discussed in this chapter can quite generally be formulated as follows. Let D be a bounded region in d-dimensional space. Since the case $d = 2$ is particularly well-known and since certain modifications are necessary for this case, we assume here that $d \geq 3$. D is assumed to be bounded by a smooth outer surface Γ and a closed set E situated strictly inside Γ. H is a class of functions harmonic in D. The problem is to give metrical conditions on E which guarantee that every function in H can be extended to a harmonic function also on E. If this is possible we say that E is removable for the class H.

THEOREM 1. *A set* E *is removable for the following classes* H *if and only if* E *has vanishing capacity for* $K(r) = r^{2-d}$:

H_1: *The class of bounded harmonic functions;*

H_2: *The class of uniformly continuous harmonic functions;*

H_3: *The class of harmonic functions with finite Dirichlet integral.*

Proof.

1) We first assume that E has positive capacity. Then there is a distribution of unit mass on E such that

$$u(x) = \int_E \frac{d\mu(y)}{|x-y|^{d-2}} \tag{1.1}$$

is bounded and even (considering if necessary a suitable restriction of μ) uniformly continuous. Since $u(x)$ is non-constant, it is an example for H_1 and H_2. To see that $u(x)$ also belongs to H_3 we observe that

$$|\text{grad } u| \leq C_1 \int_E \frac{d\mu(y)}{|x-y|^{d-1}}$$

and hence

$$\int_D |\text{grad } u|^2 dx \leq C_1^2 \iint_{E\,E} d\mu(y)\, d\mu(z) \int_{-\infty}^{\infty} \frac{dx}{|x-y|^{d-1}\,|x-z|^{d-1}} =$$

$$= C_2 \iint_{E\,E} \frac{d\mu(y)\, d\mu(z)}{|x-z|^{d-2}} < \infty .$$

2) Assume that E has vanishing capacity and construct $E_1 \supset E_2 \supset \ldots \to E$, where E_n consists of a finite number of smooth surfaces. The corresponding domains bounded by Γ and E_n are denoted D_n.

Let $u(x)$ belong to H_1 and construct $u_1(x)$ harmonic inside a suitable surface Γ_1 inside Γ such that $u_1(x) = u(x)$ on Γ_1. We consider $v(x) = u(x) - u_1(x)$ and shall prove that $v(x) \equiv 0$ inside Γ_1, which will be denoted D'.

We form the equilibrium potential of E_n

$$p_n(x) = \int_{E_n} \frac{d\mu_n(y)}{|x-y|^{d-2}} \quad (= V_n \text{ on } E_n)$$

where by assumption $V_n \to \infty$. By the maximum principle, there is a constant C independent of n so that

$$v(x) \leq C\, V_n^{-1}\, p_n(x)$$

and since $V_n^{-1}\, p_n(x) \to 0$ in D', $v(x) \leq 0$. In the same manner we prove $v(x) \geq 0$ and so $v(x) \equiv 0$.

If $u(x)$ belongs to H_3 we construct $v(x)$ and $p_i(x)$ as above. Let $\psi(t)$ be a twice continuously differentiable function defined for all real numbers t satisfying the condition

$$\psi''(t) \equiv 0, \quad |t| \text{ large.} \tag{1.2}$$

We observe that $\Delta\psi(v(x)) = \psi''(v)\,|\text{grad } v|^2$. By Green's formula we find, if $d\sigma$ denotes the surface element and n the outer normal,

$$\int_{\partial D_i} \psi(v) \frac{\partial p_i}{\partial n}\, d\sigma = -\int_{D_i} (p_i - V_i)\psi''(v)|\text{grad } v|^2 dx + \psi'(0) \int_{\Gamma_1} (p_i - V_i) \frac{\partial v}{\partial n}\, d\sigma$$

$$= \int_{D_i} \psi'(v)\,(\text{grad } p_i,\ \text{grad } v)\, dx.$$

We divide the last two expressions in the above equality by V_i and let $i \to \infty$. Since

$$\int_{D_i} |\text{grad } p_i|^2 dx = O(V_i), \quad i \to \infty,$$

which is proved as in part 1) above and since $|\psi'|$ is bounded, it follows from Schwarz's inequality that the second expression tends to zero. Since also

$$\psi'(0) \int_{\Gamma_1} (p_i V_i^{-1} - 1) \frac{\partial v}{\partial n}\, d\sigma \to -\psi'(0) \int_{\Gamma_1} \frac{\partial v}{\partial n}\, d\sigma = -C\,\psi'(0)$$

we have

$$\int_{D'} \psi''(v)\,|\text{grad } v|^2 dx = \int_{-\infty}^{\infty} \psi''(t)\, d\mu(t) = C\,\psi'(0) \tag{1.3}$$

for all functions ψ satisfying (1.2) if $\mu(a) = \iint_{v<a} |\text{grad } v|^2 dx$. (1.3) implies $\mu(a) = C_1 a$, $a > 0$, $= C_2 a$, $a < 0$. Unless $C_1 = C_2 = 0$ this contradicts $v \in H_3$. Hence $v \equiv \text{Constant} = 0$.

2. Theorem 1 shows that if the set E has dimension $> d-2$ (i.e.

$\Lambda_\alpha(E) > 0$ for some $\alpha > d-2$) there is a uniformly continuous function u with singularities on E. On the other hand we can construct very regular harmonic functions outside a smooth surface, i.e. a set of dimension $d-1$. A result connecting these two facts is given in the following theorem.

THEOREM 2. *A set E is removable for the class H_α of harmonic functions satisfying a Lipschitz condition of order α, $0 < \alpha < 1$,*

$$(2.1) \qquad |u(x) - u(x')| \leq \text{Const.}\ |x - x'|^\alpha, \quad x, x' \in D,$$

if and only if $\Lambda_{d-2+\alpha}(E) = 0$.

Proof.

1) We first assume that $\Lambda_{d-2+\alpha}(E) > 0$. By Theorem II.1, there is a distribution μ of unit mass on E such that

$$\mu(S) \leq C\, r^{d-2+\alpha}$$

for all spheres of radius r. We shall prove that

$$u(x) = \int_E \frac{d\mu(y)}{|x-y|^{d-2}}$$

satisfies (2.1). We define $\mu(r, x) = \mu(\{y \mid |y - x| < r\})$ and find for x, $x' \in D$, $|x - x'| = \delta$,

$$u(x) - u(x') = \int_0^\infty r^{2-d}\, d\mu(r, x) - \int_0^\infty r^{2-d}\, d\mu(r, x') =$$

$$= (d-2) \int_0^\infty (\mu(r, x) - \mu(r, x'))r^{1-d}\, dr$$

$$\leq C_1 \int_0^{2\delta} r^{d-2+\alpha} r^{1-d}\, dr + (d-2) \int_{2\delta}^\infty (\mu(r, x) - \mu(r-\delta, x))r^{1-d} dr$$

$$< C_2\, \delta^\alpha + (d-2) \int_\delta^\infty \mu(r, x)(r^{1-d} - (r+\delta)^{1-d})\, dr$$

$$< C_2\,\delta^{a} + C_3 \int_{\delta}^{\infty} \frac{r^{d-2+a}\delta}{r^{d}}\,dr = C_4\,\delta^{a}.$$

Since x and x' can be interchanged we have proved (2.1).

2) We now assume that $\Lambda_{d-2+a}(E) = 0$ and that $u(x)$ satisfies (2.1). We construct $v(x)$ as above and shall prove $v(x) \equiv 0$.

We can cover E by n closed spheres S_ν,

$$S_\nu: \quad |x - x_\nu| \le r_\nu$$

such that

$$\Sigma\, r_\nu^{d-2+a} \le \varepsilon.$$

We assume that ε has its smallest value when the number of spheres is $\le n$. In the proof we shall also use the expanded spheres

$$S_\nu(t): \quad |x - x_\nu| \le r_\nu t, \qquad 1 \le t \le 3.$$

For $t > 1$ every point of E is strictly inside $\cup S_\nu(t) = \Sigma(t)$. The part of the boundary $\partial\Sigma(t)$ of $\Sigma(t)$ which is boundary of the unbounded component of the complement of $\Sigma(t)$ is denoted $\sigma(t) = \cup \sigma_\nu(t)$, where $\sigma_\nu(t)$ is $\sigma(t) \cap \partial S_\nu(t)$. Clearly $\sigma(t)$ does not meet E.

By Green's formula, we have, if the function ψ is defined by the first sign of equality below, $t > 1$,

$$(2.2) \qquad \psi(t) = \int_{D'\Sigma(t)} |\mathrm{grad}\, v|^2\, dx = \int_{\sigma(t)} v \frac{\partial v}{\partial n}\, d\sigma = \frac{1}{2} \int_{\sigma(t)} \frac{\partial v^2}{\partial n}\, d\sigma.$$

If $v \not\equiv$ constant, $\psi(t)$ is bounded from below in $1 < t \le 3$, if ε is small enough. We rewrite (2.2) introducing the unit sphere U. Points on U are denoted ξ and its area element dA_ξ. The part of U for which $x_\nu + t r_\nu \xi \in \sigma_\nu(t)$ is called $a_\nu(t)$. Integrating (2.2) and using these notations we find

$$(2.3)\quad -2\int_2^3 \psi(t)\, t^{1-d}\, dt = \sum_{\nu=1}^{n} r_\nu^{d-2} \int_2^3 dt \int_{a_\nu(t)} \frac{\partial}{\partial t} v^2(x_\nu + r_\nu t\xi)\, dA_\xi .$$

In each term to the right of (2.3) we shall now interchange the order of integration. We must then study for ξ fixed for which values of t a certain ray $x_\nu + r_\nu t\xi$ belongs to $\sigma_\nu(t)$. We distinguish four cases, the first two of which are trivial.

(a). $x_\nu + r_\nu t\xi \notin \sigma_\nu(t)$, $2 \le t \le 3$. For such a ξ we get 0 as contribution to (2.3).

(b). $x_\nu + r_\nu t\xi \in \sigma_\nu(t)$, $2 \le t \le 3$. We can evaluate the t-integration and get the contribution

$$v^2(x_\nu + 3r_\nu\xi) - v^2(x_\nu + 2r_\nu\xi) = O(r_\nu^{\alpha}).$$

(c). The remaining possibility is:
$x_\nu + r_\nu t\xi \in \sigma_\nu(t)$, $\tau_i \le t \le \tau_i'$, $i = 0, 1, 2, \ldots, m$, $2 \le \tau_0 < \tau_0' < \tau_1 < \ldots < \tau_m' \le 3$. For every τ_i', $i < m$, there is an index $\mu \ne \nu$ so that $x_\nu + r_\nu \tau_i'\xi \in \sigma_\mu(\tau_i')$. We here have two essentially different cases.

(c 1). $r_\mu \ge r_\nu$. If we consider the two-dimensional plane containing x_ν, x_μ and $x = x_\nu + r_\nu\tau_i'\xi$, we see that $x' = x_\nu + r_\nu t\xi$, $t > \tau_i'$, must be interior to $S_\mu(t)$ and hence $x_\nu + r_\nu t\xi \notin \sigma_\nu(t)$, $t > \tau_i'$. (c 1) can thus occur only if $i = m$.

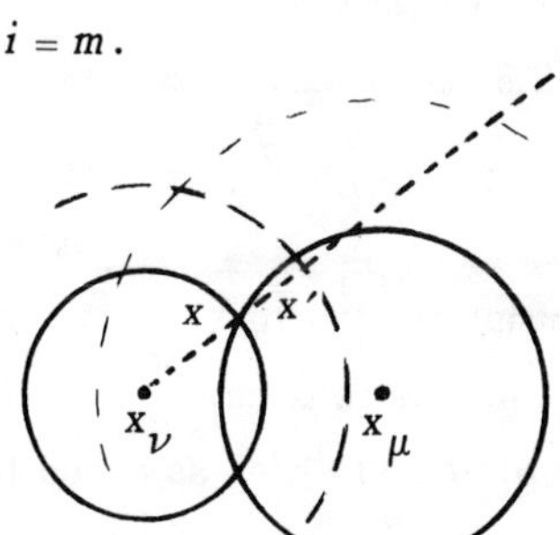

(c 2) We may thus assume $r_\mu \le r_\nu$. We first observe that $x_\nu + r_\nu t\xi$, $2 \le t \le 3$, belongs to a certain sphere $S_\mu(t)$ in a t-interval and that its length is $\le 6\, r_\mu r_\nu^{-1}$. We now consider an interval (τ_i', τ_{i+1}). The corresponding spheres $S_\mu(t)$ are $\mu = \mu_1, \ldots, \mu_k$. We can write if $\phi(t) = v^2(x_\nu + r_\nu t\xi)$,

(2.4) $$\phi(\tau_{i+1}) - \phi(\tau_i') = \sum_{j=1}^{k} (\phi(s_{j+1}) - \phi(s_j))$$

where each pair s_j, s_{j+1} belongs to one $S_{\mu_\ell}(3)$. Hence

(2.5) $$|\phi(\tau_{i+1}) - \phi(\tau_i')| \leq \Sigma \cdot C \cdot r_\nu^{\alpha} \, |s_{j+1} - s_j|^{\alpha} \leq C \cdot 6^{\alpha} \sum_{j=1}^{k} r_{\mu_j}^{\alpha}.$$

We now evaluate the t-integral of the ν:th term in (2.3) and find

$$\sum_0^m (\phi(\tau_i') - \phi(\tau_i)).$$

If we add the relations (2.4) for $i = 0, 1, \ldots, m-1$, and use (2.5) we get the estimate

(2.6) $$O(r_\nu^{\alpha}) + O(\Sigma^1 \, r_\mu^{\alpha}),$$

where Σ^1 indicates that the summation is extended over those μ such that $x_\nu + r_\nu t\xi$, $t \leq 3$, meets $S_\mu(3)$.

We consider the estimate (2.6) for different points $\xi \in U$. The "area" of U for which $x_\nu + r_\nu t\xi \in S_\mu(3)$ for some t is $O(r_\mu^{d-1} r_\nu^{1-d})$. The total ν:th term in (2.3) is thus

(2.7) $$O(r_\nu^{d-2+\alpha}) + O(r_\nu^{-1} \, \Sigma^2 \, r_\mu^{d-1+\alpha}),$$

where Σ^2 indicates summation over those μ for which $S_\mu(3) \cap S_\nu(3) \neq \phi$ and $r_\mu \leq r_\nu$. The last relations imply $S_\mu(1) \subset S_\nu(7)$. Since the covering by the spheres $S_\nu = S_\nu(1)$ was assumed to be minimal we have

$$\Sigma^2 \, r_\mu^{d-2+\alpha} \leq 7^{d-2+\alpha} \, r_\nu^{d-2+\alpha}.$$

If we use this and $r_\mu \leq r_\nu$ in (2.7), we find that the ν:th term in

(2.3) $= O(r_\nu^{d-2+\alpha})$ and so

$$\int_2^3 \psi(t)\, t^{1-d}\, dt \le \text{Const} \sum_1^n r_\nu^{d-2+\alpha} \le \text{Const} \cdot \varepsilon .$$

Hence $\psi(t)$ cannot be bounded from below and so $v(x) \equiv$ constant, and then $v \equiv 0$, as was to be proved.

3. As an analogue of Theorem VI.1 we consider the spaces $H^p(D)$, $p \ge 1$, D a bounded region, of functions u harmonic in D such that

$$\int_D |u(x)|^p\, dx < \infty . \tag{3.1}$$

The problem of removable singularities for the spaces H^p is only interesting if $d \ge 3$ since for $d = 2$ $\log |x|$ belongs to H^p for all p. We shall prove the following theorem.

THEOREM 3. *Suppose that* $d \ge 3$ *and define* q *by* $p^{-1} + q^{-1} = 1$ *and assume* $d > 2q$. *Then if* $\Lambda_{d-2q}(E) < \infty$, E *is removable for the space* $H^p(D)$. *If* $\Lambda_\alpha(E) > 0$ *for some* $\alpha > d - 2q$, E *is not removable for* $H^p(D)$.

Remark. If we as usual let $p = \infty$ correspond to bounded functions, the above result is the best possible criterion in terms of Λ_α as shown by Theorem 1 and the results of Chapter III. —We also note that $|x|^{2-d}$ belongs to $H^p(D)$ if $p < d(d-2)^{-1}$ and $p = d(d-2)^{-1}$ corresponds to $d = 2q$.

For the proof of Theorem 3 we need the following

LEMMA. *Let* $\bigcup_{\nu=1}^n S_\nu = S$ *be a covering of* E *with spheres of radii* $< \varepsilon$ *and let* x_0 *be a fixed point of* D, $x_0 \in D - S$. *Let* $d\omega(x)$ *be the harmonic measure of a surface element* dS *of* S *of area* $d\sigma$

and situated on S_ν *of radius* r_ν. *Then there is a constant* C *only depending on* D, E *and* x_0, *such that as* $\varepsilon \to 0$

$$d\omega \leq C\, r_\nu^{-1}\, d\sigma. \tag{3.2}$$

Proof. By the principle of extension $d\omega \leq d\omega_1$ where $d\omega_1$ is the harmonic measure of the ring domain bounded by S_ν and a sphere Σ concentric to S_ν of radius R, where R is large enough. Furthermore, if S_ν is $|x| < r_\nu$,

$$\int_{S_\nu} d\omega_1 < \frac{r_\nu^{d-2}}{|x|^{d-2}} \tag{3.3}$$

and by Harnack's principle $d\omega_1$ and $d\omega_1'$ corresponding to different rotations of dS on S_ν satisfy inequalities

$$\frac{d\omega_1}{d\omega_1'} < \text{Const.} \tag{3.4}$$

Since finally $|x|$ is bounded from below at x_0, (3.3) and (3.4) imply (3.2).

Proof of Theorem 3.

1) We first assume $\Lambda_{d-2q}(E) < \infty$ and consider u satisfying (3.1). As in the other proofs in this section we may assume that $u = 0$ on $\partial D - E$ and shall prove $u \equiv 0$.

We consider a covering $S = \bigcup_{\nu=1}^{n} S_\nu$ such that $r_\nu \leq \varepsilon$ and

$$\Sigma\, r_\nu^{d-2q} \leq 2\Lambda. \tag{3.5}$$

We also consider the extended spheres $S_\nu(t)$: $|x - x_\nu| = r_\nu t$, $1 \leq t \leq$ ≤ 2, and as in the proof of Theorem VI.1 (c) we can write

$$u(x_0) = \sum_{\nu=1}^{n} \int_{\gamma_\nu(t)} u(x)\, d\omega_t(x), \tag{3.6}$$

where $d\omega_t$ are the harmonic measures described in the lemma for $S = \cup S_\nu(t)$. We integrate (3.6) with respect to t, $1 \le t \le 2$, and use (3.2) and find

$$(3.7) \qquad |u(x_0)| \le C \sum_{\nu=1}^{n} r_\nu^{-1} \int_1^2 dt \int_{\gamma_\nu(t)} |u(x)|\, d\sigma .$$

$\gamma_\nu(t)$, $1 \le t \le 2$, covers a domain Ω_ν and these domains Ω_ν are except for the boundaries disjoint. (3.7) then yields by Hölder's inequality

$$|u(x_0)| \le C \sum_{\nu=1}^{n} r_\nu^{-2} \int_{\Omega_\nu} |u(x)|\, dx \le$$

$$\le C \sum_{\nu=1}^{n} r_\nu^{-2} \left\{ \int_{\Omega_\nu} |u(x)|^p\, dx \right\}^{\frac{1}{p}} \left\{ \int_{\Omega_\nu'} dx \right\}^{\frac{1}{q}} \le$$

$$\le C_1 \sum_{\nu=1}^{n} r_\nu^{d \cdot q^{-1} - 2} \int_{\Omega_\nu} |u(x)|^p\, dx^{\frac{1}{p}} \le$$

$$\le C_1 \left\{ \int_{\Omega} |u(x)|^p\, dx \right\}^{\frac{1}{p}} \left\{ \sum_{\nu=1}^{n} r_\nu^{d-2q} \right\}^{\frac{1}{q}} .$$

Here $\Omega = \cup \Omega_\nu$ is a neighbourhood of E. Choosing ε small the first factor is small while the second by (3.5) is bounded. Hence $u(x_0) = 0$ as asserted.

2) We now assume $\Lambda_\alpha(E) > 0$, $\alpha > d - 2q$, and choose a distribution of unit mass on E such that $\mu(S) \le C r^\alpha$ for all spheres of radius r. We prove that

$$u(x) = \int \frac{d\mu(y)}{|x-y|^{d-2}}$$

belongs to H^p. We use the method of Theorem VI.1 and consider a step function $f(x) > 0$ such that

$$\int_D f(x)^q\, dx = 1 .$$

Define $\beta = p^{-1}\alpha + q^{-1}(d-2q)$ so that $d-2q < \beta < \alpha$ and consider

$$\phi(s) = \int_D \int_E \frac{f(x)^{q(1-s)} d\mu(y)\, dx}{|x-y|^{\beta+(d-\alpha)s}} .$$

We see that $|\phi(it)| \leq M$ and $|\phi(1+it)| \leq M$, with M independent of $f(x)$. By the maximum principle,

$$\phi(p^{-1}) = \int_D f(x)\, u(x)\, dx \leq M ,$$

which then yields

$$\int_D |u(x)|^p\, dx \leq M^p .$$

[1] AHLFORS, L.: Bounded analytic functions. M.R. *9*, 24, (1948)
[2] AHLFORS, L. and Beurling, A.: Conformal invariants and function-theoretic null-sets. M.R. *12*, 171, (1951)
[3] BESICOVITCH, A.S.: On existence of subsets of finite measure of sets of infinite measure. M.R. *14*, 28, (1953)
[4] BEURLING, A.: Ensembles exceptionnels. M.R. *1*, (1940)
[5] CALDERON, A.P.: On the behaviour of harmonic functions at the boundary. M.R. *11*, 357, (1950)
[6] CARLESON, L.: Interpolations by bounded analytic functions and the Corona problem. Ann. of Math. *76*, 547–559, (1962)
[7] CARLESON, L.: On a class of meromorphic functions and its associated exceptional sets. M.R. *11*, 427, (1950)
[8] CARLESON, L.: On null-sets for continuous analytic functions. M.R. *13*, 23, (1952)
[9] CARLESON, L.: On the connection between Hausdorff measures and capacity. M.R. *19*, 1047, (1958)
[10] CHOQUET, G.: Theory of capacities. M.R. *18*, 295, (1957)
[11] DAVIES, R.O.: Subsets of finite measure in analytic sets. M.R. *14*, 733, (1953)
[12] DENY, J.: Les potentiels d'énergie finie. M.R. *12*, 98, (1951)
[13] ERDÖS, P. and GILLIS, J.: Note on the transfinite diameter. J. Lond. math. Soc. XII (1937)
[14] FROSTMAN, O.: Potential d'équilibre et capacité des ensembles avec quelques applications à la théorie des fonctions. Meddel. Lunds Univ. Mat. Sem. 3, (1935)
[15] GARABEDIAN, P.R.: Schwarz's lemma and the Szegö kernel function. M.R. *11*, 340, (1950)
[16] KAMETANI, S.: On some properties of Hausdorff's measure and the concept of capacity in generalized potentials. M.R. *7*, 522, (1946)
[17] KAMETANI, S.: A note on a metric property of capacity. M.R. *15*, 622, (1954)
[18] KISHI, M.: Capacities of borelian sets and the continuity of potentials. M.R. *20*, 422, (1959)
[19] KUNUGUI, K.: Etude sur la théorie du potentiel généralisé. M.R. *12*, 410, (1951)

[20] NEVANLINNA, R.: Ein Satz über offene Riemannsche Flächen. M.R. *2*, 85, (1941)

[21] OHTSUKA, M.: Capacité d'ensembles de Cantor généralisés. M.R. *19*, 541, (1958)

[22] SALEM, R. and ZYGMUND, A.: Capacity of sets and Fourier series. M.R. 7, 434, (1946)

[23] TEMKO, K.V.: Convex capacity and Fourier series. M.R. *19*, 31, (1958)

[24] TSUJI, M.: Beurling's theorem on exceptional sets. M.R. *12*, 692, (1951)

[25] TSUJI, M.: A simple proof of a theorem of Erdös and Gillis on the transfinite diameter. M.R. *18*, 650, (1957)

NOTE: To save space, references are made only to Mathematical Reviews (*M.R.*).